BASICS OF
FLEET MAINTENANCE

JOEL LEVITT

Basics of Fleet Maintenance
Joel Levitt
ISBN 978-0-9825163-4-8
HF022020

©2014-2020, Reliabilityweb, Inc.
Original © Copyright 2010 Reliabilityweb.com.
All rights reserved.
Printed in the United States of America.

This book, or any parts thereof, may not be reproduced, stored in a retrieval system, or transmitted in any form without the permission of the publisher.

Opinions expressed in this book are solely the authors and do not necessarily reflect the views of the Publisher.

Publisher: Reliabilityweb, Inc.
Page Layout and Cover Design: Patricia Serio

For information: Reliabilityweb.com
www.reliabilityweb.com
8991 Daniels Center Drive, Suite 105, Ft. Myers, FL 33912
Toll Free: 888-575-1245 | Phone: 239-333-2500
E-mail: crm@reliabilityweb.com

20 19 18 17 16 15 14 13 12 11 10

Contents

Author Bio	*vii*
Chapter 1: Introduction	1
Chapter 2: Fleet Issues In Today's Marketplace	5
Chapter 3: Fleet Management	7
Chapter 4: Lean Maintenance	11
Chapter 5: Where Are You Today?	17
Chapter 6: Basis For Making Decisions	37
Chapter 7: Maintenance Costs	57
Chapter 8: Preventive Maintenance Systems and Procedures	69
Chapter 9: Measuring Worker Productivity	93
Chapter 10: Work Standards	101
Chapter 11: Staffing	109
Chapter 12: Shop Design	115
Chapter 13: Planning and Scheduling	119
Chapter 14: MRO Inventory	131
Chapter 15: Purchasing Maintenance Parts	145
Chapter 16: Vendors	149
Chapter 17: Warranties	155
Chapter 18: Fundamental Questions About Fuel	159
Chapter 19: Tire Management	169
Chapter 20: Leasing Mobile Equipment	175
Chapter 21: Loss Prevention and Fleet Safety	179

Chapter 22: Ideas to Reduce Your Insurance Costs	183
Chapter 23: Trading Vehicles	185
Chapter 24: Vehicle and Equipment Specification	187
Chapter 25: Case Study In Alternative Use	195
Chapter 26: Vehicle Maintenance Reporting Standards	197
Chapter 27: The Computer Generated Repair Order	201
Chapter 28: Computerized Maintenance Management System	209
Chapter 29: Driver Vehicle Inspection Reports	221
Chapter 30: Using Statistics To Identify Problems	223
Chapter 31: Fleet Key Performance Indicators	229
Chapter 32: Routing Systems	235
Chapter 33: Vehicle Locating Systems	237
Chapter 34: Vehicle Identification Systems	239
Chapter 35: Wireless	241
Chapter 36: Fifty Notes to Take With You	243

Author Bio

Since 1980, Mr. Joel Levitt has been the President of Springfield Resources, maintenance consultants in a wide variety of industries including transport, Port, Private Fleets, oil, airports, Universities, hospitals, high tech manufacturing, school systems, government, etc.

Mr. Levitt is a leading maintenance trainer throughout the US, Canada, Europe and Asia. He has trained over 15,000 maintenance professionals from 20 countries in 500+ sessions. 98% of those individuals have rated the training as very good or excellent.

Prior to SRC, Mr. Levitt was a Senior Consultant at Computer Cost Control Corporation, a Computerized Maintenance Management Systems company for fleets. His duties included assisting the president in designing and marketing computerized fleet maintenance management systems to organizations including FedEx, United Airlines, JFK Airport, BFI, etc. He also provided software for tire management, routing, fuel management and dispatch.

Prior to that, as President of Springfield Controls, he installed and serviced Fuel Management systems for private fleets. At the same time, Mr. Levitt designed, installed and serviced fuel automation systems with rack control, accounting, and inventory control for BP North America's 30,000-barrel/day-oil terminal. He wrote the standard for the American Rail Road Association Locomotive Refueling.

Mr. Levitt has written seven other books on maintenance management, chapters in two other books and over seventy articles in the trade press. He has been a columnist for Fleet Maintenance Magazine since 2004.

Additional titles by Joel Levitt:
- Complete Guide to Predictive and Preventive Maintenance
- The Handbook of Maintenance Management, 2nd Edition
- Lean Maintenance
- Maintenance Planning, Scheduling, & Coordination (with Nyman)
- Managing Factory Maintenance, 2nd Edition
- Managing Maintenance Shutdowns and Outages
- TPM Reloaded: Total Productive Maintenance

Introduction

The field of Fleet Maintenance is unique when compared to maintenance in general. This uniqueness comes from the sheer quantities of maintainable units involved. In 2006, the U.S. Department of Transportation reported there were 250,851,833 cars and light trucks on the road. Since 1972, that number has exceeded the number of licensed drivers. This number does not include truck, combination (heavy trucks), construction equipment, or any other type of fleets (locomotives, tugboats, airplanes, etc.).

These numbers drive the Fleet Maintenance world in a variety of ways. Maintenance workers in this field can be quite specialized. There are transmission specialists, tire specialists, diesel specialists and others. You don't commonly find this kind of specialization in other areas of maintenance. When you do (such as in copy machine repair) you also find large populations of maintainable units, much like fleets.

As these fleets began to grow, a large and well organized parts distribution business developed around spare parts. Standard inventory models were applied to managing spare parts rooms. All parts were charged to units and all parts (coming and going) were accounted for. When computers were applied to the early fleets, the storerooms were also computerized. One unique development is the way in which parts were numbered. Many truck companies (this is particularly true in the U.S., less so in Europe) sell the same engines and drive line components under their own name, applying their own part numbers. The problem then becomes that the same part may now have several part numbers. Cross reference books, and now web sites, were developed which allowed simplification of the stocking problem across heterogeneous fleets.

These large populations allowed for the use of statistics in maintenance. Since a large fleet might have 1000 units or more (the U.S. Post Office fleet was purported to be 260,000 units) statistics could be applied to analyses. Different distributions could be fitted to the mass of data to come up with defensible conclusions.

One of the results of the size of the fleet market (and the opportunity it offered) is that it became computerized before other maintenance areas. Fleet Maintenance computer systems developed early and (using the most powerful computers of the day) became quite sophisticated, including early development and adoption of the CMMS in the fleet world.

In 1968 (very soon after the first primitive computer system to aid maintenance), the American Trucking Association began to codify data, including repairs, components, unit information, locations, etc., into the Vehicle Maintenance Reporting Standard (VMRS) still in use today. This standard was a huge benefit for the developers of CMMS in the Fleet Maintenance field since much of the needed information had already been developed (unlike CMMS developers in the factory or building maintenance markets). These standards spurred development. Concepts developed twenty years ago in fleet systems are just now appearing in the current crop of CMMS in other markets.

The development of codes for all repairs, and standardized ways of discussing work accomplished, was a huge advantage to labor management. Fleet systems quickly developed historic labor standards based on individual locations with unique skill sets, tools, and topography (layout). This simplified supervision since it was immediately obvious when a job was taking too long.

Another benefit from the rapid development of competent CMMS was the information it supplied to decision makers. Repair histories became readily available to identify costs, frequencies, and performance of component systems since purchase. The computers could easily aggregate the data across many vehicles of a similar size, design and use (called a class of vehicles), and come up with great insights into what was best for a particular fleet. These accurate costs over time encouraged managers to make better buying decisions. Imagine having comparison data of repairs to like units in like service at your fingertips when shopping for new units.

This field is also unique from a business point of view. There are the OEMs (Original Equipment Manufacturers) and the dealers. Unlike other businesses, dealers are independent businesses. There was an incentive to develop detailed repair manuals (for training and quality control) and to accurately time repairs (to define warranty reimbursements).

The result is that Fleet Maintenance has excellent manuals that are commonly available, along with databases of repair times. This means the maintenance shops have standards and can (relatively) easily institute incentive programs. In fact, some car dealer repair departments run wholly on labor standards where the mechanic gets paid the book time for the job times his/her personal labor rate. Good mechanics can make quite a bit of money this way. When you combine the labor manuals with the ease of generating historical standards, you have a winning combination for shop management and scheduling.

Other special conditions have to do with the large numbers of units, which allows for standardizing of data. They include simplified shop scheduling, ease in predicting the number of mechanics needed to support a fleet and the impact of changing the size of the fleet, and the quantity of fleet specific skills training in almost every major metropolitan area.

Fleet Issues In Today's Marketplace

As service providers to our companies (or other companies), we face a number of issues. Additionally, many of us manage private fleets in organizations where the mission is to provide a product or service and it is our responsibility to ensure the smooth operation of the "getting it there" part of the supply chain.

Following is a set of issues developed from a recent Fleet Maintenance class. Perhaps you face many of these same issues along with some additional ones of your own.

- Price of fuel
- Volatility in the price of fuel
- Not enough work
- Uncertain economic environment
- Not enough (or too much) rolling stock
- Not enough (or too much) work for shop
- Unknown costs to own or operate vehicles
- Abuse of vehicles by drivers
- Difficulty in finding qualified people
- Unknown if vehicles are properly specified
- Information needed re: rebuild vs. new in components and vehicles
- Low shop labor productivity
- Drivers/mechanics not passing drug test
- Too many emergencies
- Need for a PM system that works
- Scheduling the shop
- Keeping an old fleet on the road without breakdowns
- Problems with collecting accurate data
- Security problems
- Insufficient information/time/etc. to prepare budget
- More information needed on the lease/buy issue

- Information needed on rebuilding in-house vs. outside
- Frequent stock-outs of parts
- Bad shop layout
- How to compare MPG
- Tires
- Shopping for a computer system
- Locating vehicles

Fleet Management

Let's start at the beginning:
- What are the overall responsibilities of the fleet manager?
- Why should an organization undertake the job of managing their fleet?
- What are the alternatives to fleet management?
- How does Fleet Maintenance affect quality of life issues?

3.1 Overall Responsibilities of the Fleet Manager

Depending on the organization, the fleet manager has different responsibilities. In some organizations, the fleet manager may only be responsible for the maintenance of the equipment while others are responsible for equipment purchasing, fuel taxes, route management, etc. This book will encourage you to take a larger view of fleet operations. Efficient operation of the fleet requires that management have the ability to impact all fleet responsibilities. The following is a partial list of fleet manager responsibilities:

- Corporate information systems (Enterprise systems)
- Customer services
- DOT
- Drivers
- Environmental, safety and health
- Fixed asset management
- Fleet availability
- Fleet information system (also CMMS)
- Fuel
- Fuel tax
- Government requirements
- Life cycle costing
- Maintenance

- Mechanics
- Other transportation systems
- Parts
- Permits and licenses
- Route management
- Safety
- Security
- Shop management
- Staff deployment
- Tires
- Vehicle dispatch
- Warehousing
- Warranty recovery

These issues all apply to the discussion of fleet management. This book concentrates on ways to look at all aspects of the fleet operation to help manage resources and avoid costs.

3.2 Why Manage the Fleet?

You can manage (try to control what happens and plan) your fleet or you can just let things happen (repair what breaks after it breaks). This debate (to manage or not to manage) has raged since the beginning of the automotive era.

For example, in October of 1926, an article titled, "Maintenance Costs Cut Through Regular Inspections", appeared in Bus Transportation. This article argued the advantages of periodic inspections versus breakdown mode maintenance. Earlier references include articles in the magazine of the Society of Automotive Engineers. For instance, "Care and Maintenance of Motor Trucks", was featured in the April 1921 issue.

Each fleet has to deal with this issue in their unique way and the ideal level of management for each fleet changes over time. There are excellent reasons to manage the fleet, including many of the following:
- Cost reduction
  - A well-managed fleet costs less to run in the long term.
- Cost control
  - A well-managed fleet's cost of operation will vary in a controlled way rather than random breakdowns driving yearly costs.

- Service
  - Good management provides a better level of service to fleet users or outside customers. Service is defined according to the needs of the user, i.e., on time, reliable, undamaged, clean, etc.
- Employee needs
  - A managed fleet is a good place to work. Operators have well-maintained, safe equipment to operate. Mechanics have a productive, managed atmosphere that rewards a longer-term view of the fleet operation.
- Safety and environmental protection
- Quality of life
  - The quality of life for all involved is higher. There are fewer middle-of-the-night breakdowns and equipment failures and more time to spend on other activities.

3.3 Alternatives to Fleet Management

We live in a society where almost any service is available for a price. If you prefer to concentrate on your main businesses rather than your fleet, you have that option. But you DON'T have the option to ignore your fleet. The costs are far too high. From a cost effectiveness point of view, you have a few alternatives. You can manage your fleet or do one of the following:

- Spin off the fleet as a separate profit center. Hire someone to run the separate fleet company and give them an incentive based on the new company's profit. Because of the advantages of this method, many organizations set up fleet subsidiaries. These subsidiaries are especially useful if the main organization is in some other business such as education, wholesale building products, beverage bottling, etc.
- Turn your fleet "problem" over to a full-service leasing company. The Ryder's and Leaseway's (in the USA alone there exist about 15,000 smaller lessors) would be happy to make a profit by managing your fleet. They can lease equipment and will even lease drivers or trips. The leasing companies can design a service that expands to fit your peak periods and contracts when you are slow. Until recently, even large fleets (such as Sear's Signal Delivery fleet) were operated under a full-service lease. (This option will be discussed further in the next section.)
- Sell your trucks to the drivers and make them owner-operators. They become responsible for the management of their own small (one unit) fleet. Owner-operator fleets are very popular among

common carriers. The owner-operators allow the carriers to expand and contract to fit market conditions.
- Bring in a service company who will charge you per mile, trip, or ton. This company makes a profit by managing your fleet. This is similar to full-service leasing where all fleet activity is transferred including drivers, garages, and tools. Many successful current fleet managers would jump at the chance to "own" a transportation company.

3.4   Fleet Maintenance and Quality of Life Issues

One of the most important reasons to manage your fleet is to improve the quality of life for yourself and your employees. Quality of life means different things to different people. In general, good quality of life means the following:

- The organization is operated in a consistent manner so that all employees feel as if they are treated fairly. The organization is stable and satisfies the employees' security concerns. If the organization is not stable, employees are told the truth.
- Employees are viewed as capable contributors to the organization's well being. It means changing the traditional view of the mechanic as a grease monkey. A quality organization views mechanics as highly skilled professionals and is willing to invest in their training credentials.
- Employees have the time, skills, and resources to do a good job and get the feedback they deserve. The organization is supportive of its managers, staff, and mechanics, educating them and advancing their skills.
- The organization supports a healthy home life. This means that managers are not expected to work 70 to 80 hours per week (except for infrequent emergencies). Emergency calls in the middle of the night are exceptions, as are missed family events.
- The organization is operated so that healthy relationships between the people that work there are encouraged. Collaboration is valued over competition.
- The organization maintains a physically healthy work place that is free from hazardous materials, or, when hazardous materials are present, they are handled properly (to the best of current knowledge). Safety of the employees, community, and the environment is a core mission of the organization.

Lean Maintenance

Lean Manufacturing is based on the work of Dr. Shigeo Shingo at Toyota. He states, **"Lean is an all out war against waste from both manufacturing inefficiencies and under-utilization of people."** This same attitude can be used to reduce waste in fleets.

In some maintenance facilities, you see the waste from the moment you walk into the shop. It looks like debris and dirt on the floors, in the corners, and under the benches. It looks like vehicles backed up waiting for parts or an open bay. It sounds like complaints of being out of (you can fill in the blank) or being unable to find (fill in).

Lean maintenance elimination can be lead from the top down but must be executed by the rank and file. Why? In years past, the supervisor, manager, or superintendent was the one responsible for ideas for improvements. Maintenance people were the "hands" hired to do what they were told. Today organizations are lean and mean. In order to succeed, you need all the capabilities of all the people in that organization. The downsizing craze has left anyone in a managerial role with too many tasks and too little time. There is no one left to cut costs!

One axiom of Lean Maintenance is the wisdom of the maintenance, yard jockeys, tire guys, and cleaning people. These men and women are in daily contact with the facility and see firsthand the wasted, unused, and improperly used resources. If you want to implement waste cutting make it a routine part of every job assignment!

Also, this is the fun stuff! The most stimulating and interesting part of maintenance, for most workers, is solving problems. It's also a team activity. Ask any one of your maintenance people where they get the most satisfaction from their job. Inevitably, the answer will be when fixing the "big" breakdown, really helping a customer, or permanently solving a problem.

Lean Maintenance can be as simple as improving the positioning and quantity of wheeled oil drain pans, or as complicated as evaluating two different air compressors in the same application (with initial cost, availability, energy costs, reliability, repair costs, availability of spare parts, downtime, and downstream impacts to consider).

Where do you look for waste cutting? Everywhere you can imagine and in every activity you do. There are opportunities for cutting waste and making improvements in every maintenance operation. An internal study done by a major maintenance provider in Canada estimates the opportunity as follows:

Possible Savings of Maintenance Budget Dollars* (Translates to reduction of waste)	
39%	Re-engineering of equipment and maintenance improvements to equipment
26%	PM improvement and correct application of PM
27%	More extensive application of predictive maintenance
7%	Improvements in the storeroom

*This study did not include fuel.

General Resource areas to focus on:
- Labor (driver, operator, maintenance mechanic, contractor)
- Maintenance spare parts, consumables (tires, oil, antifreeze)
- Energy, fuel, other utilities (heat, light, phone)
- Truck/trailer time (reduce machine time to complete job)
- Capital (extend life of asset, cheaper asset, less equipment reduces effective capital costs)
- Management effort (reduce headaches, non-standard conditions requiring management inputs)
- Overhead

Or:

Improve quality, production, or safety, while holding resources to the same level or lower.

Another way to look at Lean Maintenance is to consider the processes and outcomes from each of the following areas (this is a starter list):
- Breakdowns
- Corrective work, CMMS and Work Order system, work requested
- Asset specification
- Shop Planning and scheduling
- PdM and PM

- Shutdowns and Projects
- RCA (root cause analysis), RCM (reliability centered maintenance)
- Stores, storage, delivery to repair bays, purchasing
- Supervision and management

The idea is to look into one of these areas and see where the waste is. Some amount of waste will always be very obvious. Create lists for each area of potential waste sources.

The lists are like gold ore in a gold mine. Some gold is just sitting in the ore for the taking (alluvial). Other gold is locked deep inside and has to be (expensively) extracted.

The first challenge is to find the alluvial gold. In Lean Maintenance, this is called low hanging fruit. Pick that fruit first! There are many potential projects for each item on your list. Use your list of potential waste sources to drive your list of projects.

4.1 Lean Projects

A good Lean project is one with quick results (alluvial), small cost that can be completed with tools and materials that are easily accessible. (Great projects may break these guidelines but this is the place to start).

Practice makes perfect. Starting with these kinds of projects, you can build on your capacity to design, justify, execute and report results so that big projects will be easier and more likely to be successful.

Characteristics of good Lean projects:
- High probability of success
- Minimal money invested
- Tools and materials easily available
- Doable within 3-4 weeks
- Shortest cycle for payback

Some of the best projects are so simple they are called "no-brainers" or "low hanging fruit".

STOP – Don't get caught by unintended consequences! When you start to focus on one project consider the unintended consequences and look for hidden problems.

Example:

Sewer authority had a problem. Truck drivers would often forget to turn the truck keys in to dispatch. This was a problem because the authority provides 24 hour coverage with the same fleet of trucks. It became a major problem when the offending driver went on vacation with the keys. Duplicate keys were available, but, occasionally, even the duplicate would end up at home.

A simple solution was devised. A 9" piece of steel rod was added to the key ring (like the broom handle piece on the lavatory key in gas stations). The idea worked perfectly. Everyone returned the keys every night.

The unintended consequence started showing up about three months later with the failure of ignition locks. It started with one or two failures but quickly became a cascade. It was discovered that the weight of the rod eventually damaged the ignition switch.

An aluminum plate the size of a candy bar (about 1/3 the weight of the rod) was substituted.

If you were to speculate on the next unintended consequence, you might worry about how the plate would act in an accident. Either way, always look at the idea and try to think of what could go wrong.

4.2   The Lean Maintenance Proposal

For smaller projects, a proposal is not needed. Just be sure to take measurements before and after and keep good notes. On larger projects, you will need to write up a proposal. Write-up the Lean Maintenance project proposal so that a non-maintenance person can understand what you plan to do, how you plan to do it, and how and where the savings will come from.

- **Planning** – Begin the process of planning your project. Assess the labor, parts, tools, access to asset, who to inform before starting, and any other elements necessary to complete the project. If your project is a study, or investigation, determine who you have to talk to and what external or internal resources will be needed.
- **Measurement** – Sometimes the most difficult part of a maintenance project is determining how to measure it. Your cost analysis will be based on the measurements that you choose.
- **Cost Benefit Analysis** – Cost analysis of all aspects of the project is essential. Your costing should be based, as much as possible, on measurements and as little as possible on guess work.

- **Team Production** – Assign team members to different aspects of the project. Assignments might include: formal project write-up, data gathering, interviews, observations, taking readings, calling vendor engineering support departments, making presentation materials, giving presentations, formal write-up of findings, etc.

4.3 The Project Itself

1. Record the data that your experiment calls for. Keep logs or journals. As results start to come in, be very careful to accurately record everything (even if it doesn't seem to apply at first glance). Don't give in to the tendency to fudge data to make the project look successful.
2. Call the vendors for additional help or for new products to solve old problems. The vendors should be very interested in the results of your project and might even provide materials or tools if they can use your data.
3. Ask yourself the question… Are the results repeatable and are measurement methods sound and complete?
4. Take a census of the size or kind of equipment that your experiment applies to. For example, in a project in a county to save money in pool chemicals, the census would be the number of pools in that county.
5. Determine: Did the project meet the financial and other objectives?

It should be no surprise that organizations run on money. The most powerful argument to bring to the table is—Will this project save the organization money?

Consider the following questions to determine your answer:
- What is the cost of the old way of doing business?
- What is the cost of the downtime? (Total Down Hrs x Rate)
- What is the cost per year?
- What is the return on investment of a projected improvement?
- How much should we spend to fix this?
- What is our investment in this asset or process?

Where Are You Today?

Following are three questionnaires and several information gathering projects to help you better understand the condition of your fleet and its management. Use this questionnaire with the Fleet audit, and the other instruments, to determine the overall status of your fleet and to uncover areas for improvement.

The advantages of knowing, in detail, the condition of your fleet before you start are manifold. Improving a fleet is like a journey. Knowing accurately where you are will make the planning of the journey more effective and increase the probability that you'll set off in the right direction.

As you do the questionnaires and information gathering projects resist the tendency to "already know the answer". Imagine you are new to the fleet and really want to know the answers. Other people's perceptions are different than yours and could be very valuable to your deeper understanding of the fleet (even if you are convinced that they are wrong).

5.1   Fleet Maintenance Fitness Questionnaire

This questionnaire can be used to:
- Find out how you are doing in your users' eyes.
- Teach others good fleet management practices.
- Identify areas that need attention.

How to Use the Fleet Fitness Questionnaire to Your Advantage
- Have your own maintenance staff fill out the questionnaire and ask them where you are today. You may be surprised at the response. Either way, it defines the language and sets the stage for further discussions.
- Give the questionnaire to Upper Management with the statement that this is what good maintenance is about and you would like their opinion as to how you are doing.
- Give the questionnaire to your users with the statement that these are your goals and how do they think you are doing.

- Feel free to modify the questions to reflect your organization and your language.

Rating Each Question's Priority
- Go through the questionnaire. Add your priority to each question using a 1–9 scale. Review your grades for each section. Use the lowest grades as a guide to future projects in those areas.
  ◇ Use a 1–9 rating scale.
  ◇ Consider only those items important to your organization.
  ◇ Compile a separate list of the important items only, in priority order.
  ◇ This list can be used as an input to your planning or pre-planning brainstorming session or added to your Project Action Plan.
  ◇ Score so that you follow the percentage. In other words, if you follow the question 50% of the time give yourself the points in the "less than 60%" column.

	INITIATION AND AUTHORIZATION OF WORK						
Q#	Do you have...?	NO	>20%	>40%	>60%	>80%	100%
1	A written formal work order (Repair Order – RO) system	0	5	10	20	35	45
2	A printed (or computer generated) Repair Order (RO)	0	5	10	15	20	20
3	A written Procedure for your RO system	0	0	0	5	10	15
4	A single person or unit responsible for screening and/or prep of all RO	0	0	10	15	20	25
5	A formally designated and trained group that can request Maintenance Service	0	0	0	5	10	10
6	All work identified by repair reason: PM, Re-build, Accident, Emergency	0	5	10	15	20	20
7	A guideline that extra authorization is required for special jobs in contrast to normal repairs	0	0	0	5	5	10
8	A guideline for a "Reasonable Date Wanted" on all RO with restrictions against the use of ASAP, AT ONCE, HOT, etc.	0	5	5	10	15	20
	Total Points (165 possible)						

Where Are You Today?

PLANNING AND SCHEDULING

Q#	Do you have...?	NO	>20%	>40%	>60%	>80%	100%
1	Historical performance standards for common repairs (actual time)	0	5	10	15	20	25
2	Periodic issuance of Earned Hour or Productivity reports	0	0	0	0	30	40
3	A productivity incentive system	0	0	5	5	10	10
4	Any feedback of a job's status before it is completed	0	0	5	15	25	30
5	An up-to-date plan for any projects such as major re-builds, special jobs, or refurbishments, with start & end dates, hours, etc.	0	5	15	20	25	25
6	Reviews of the project plan by the Transportation Mgr and other top Mgr's on a weekly basis	0	0	0	0	10	20
7	PM inspections and service done on schedule and not delayed due to large jobs or lack of labor	0	0	5	10	15	20
8	Jobs completed on time and in line with schedule and promises made	0	0	5	10	15	25
9	A process (and use it) to trend backlog to support crew size changes	0	0	5	10	15	20
10	Data (and use it) to predict scheduled overtime and use of outside shops	0	0	0	5	15	20
11	A good idea of the effect in hrs/month of changes in fleet size or mix (or after an acquisition)	0	0	0	5	10	15
	Total Points (250 possible)						

PREVENTIVE MAINTENANCE

Q#	Do you have...?	NO	>20%	>40%	>60%	>80%	100%
1	High PM compliance (>90% PMs are done in the week they are due)	0	0	0	0	5	10
2	Unique unit numbers for all units and use those unit numbers on all ROs	0	0	0	5	10	15
3	Repair history readily available to identify costs, frequencies, and component systems since equipment was purchased	0	5	15	25	35	40
4	Comparison data of repairs to like units in like service	0	5	10	15	20	20
5	Repair budgets for major units	0	0	0	5	5	10
6	PM mechanics generate ROs immediately after detecting conditions that should be corrected	0	5	10	15	20	25

PREVENTIVE MAINTENANCE (cont.)

Q#							
7	A tickler file, computer system, or some other method to automatically generate PM inspection orders when they are due	0	0	0	5	10	15
8	Mechanics assigned to PM on a full time basis (not interrupted by breakdowns or other work)	0	0	5	10	15	15
9	All PMs rationalized so that statutory PMs are incorporated into regular PMs to avoid duplication	0	0	0	5	10	15
10	Special training for PM mechanics in diagnosis and Predictive Maintenance	0	5	10	15	20	25
	Total Points (190 possible)						

STORES AND PARTS

Q#	Do you have...?	NO	>20%	>40%	>60%	>80%	100%
1	Maintenance responsible for control of maintenance stores	0	0	0	0	5	5
2	Store requisition for special parts tied to RO	0	0	0	5	10	15
3	All parts issued charged to units	0	0	5	10	15	20
4	An annual physical inventory and review with elimination of obsolete parts	0	0	5	10	15	20
5	A shortage (last physical inventory 0%, 1%, 2%, 3%, 4%, 5% or greater)	0 (>5%)	5 (<5%)	10 (<4%)	15 (<3%)	20 (<2%)	25 (<1%)
6	Controlled stock levels (Re-order points, min-max levels)	0	0	5	10	15	25
7	A parts catalog which includes a cross-reference	0	0	0	5	10	15
8	Units down awaiting parts (% of time)	20	15	10	5	0	0
9	The stores system identifies make, model, and where the part is used	0	0	0	5	10	15
10	A functioning warranty management system for parts that fail before the warranty is used up	0	0	5	5	10	20
	Total Points (180 possible)						

MAINTENANCE ADMINISTRATION

Q#	Do you have...?	NO	>20%	>40%	>60%	>80%	100%
1	An organizational chart	0	0	0	0	5	5
2	Adequate planning and clerical staff	0	5	10	15	20	20
3	A job time-keeping system to identify and account for all payroll hours against units or other assigned tasks	0	0	10	20	25	30

	MAINTENANCE ADMINISTRATION (cont.)						
4	A regular report showing labor hrs for PM, Emergency, scheduled repair and other activities	0	0	0	0	10	20
5	Regular meetings with user departments	0	0	0	0	15	20
6	The Maintenance Department head report to the VP of Transportation or Operations or Plant Manager	0	0	0	0	0	15
7	Regular identification, and review, of repeat repairs with an eye towards solving them permanently	0	0	5	10	15	20
8	A warranty management system for all vehicle warranties	0	0	5	5	10	20
	Total Points (150 possible)						

FLEET MAINTENANCE FITNESS QUESTIONNAIRE SCORE SHEET		
Section	Your Score/Total Score	Your Grade
Initiation and Authorization	/165	
Planning and Scheduling	/250	
Preventive Maintenance	/190	
Stores and Parts	/180	
Maintenance Administration	/150	
Overall Total	**/1000**	

5.2 Fleet Shop Audit

Take a quick look at your current Fleet Maintenance facility. This audit is to be conducted immediately, and annually thereafter. This audit can be used, in conjunction with the Fleet Maintenance fitness questionnaire, to determine if proper systems and controls are in place.

The author would like to thank Ron Turley, a leading expert in the field, for this audit form.

FLEET SHOP AUDIT FORM		
Priority	Question	Findings
	Review 10 random repair orders for completeness and accuracy.	
	Driver reports are reviewed and corrective work is done in a timely manner. Check 10 random reports.	

FLEET SHOP AUDIT FORM (cont.)

Priority	Question	Findings
	Written RO is in evidence for all except genuine emergency repairs. Check this immediately for all bays.	
	Flat rates (standards of some kind) for all recurring jobs written on RO. Check all RO open on shop floor now.	
	Check to see if one day of work is planned for each mechanic at least ½ day in advance.	
	Is the Maintenance Schedule visible to all mechanics?	
	Check fluid levels and batteries on 10 random vehicles.	
	Pull 10 PM sheets and verify that they are complete and correct. Track any Corrective items and verify a RO was closed on each corrective item in a reasonable time.	
	Is the building clean, including floors, walls, eating areas, lavatories, and lighting fixtures (adequate light)?	
	Is the outside yard free from debris, oil, diesel and metal objects (nuts, bolts, nails, small parts?)	
	Is there logic to where vehicles are parked in the yard? Can you see it?	
	Is the Maintenance office clean and straightened up (consistent with use)?	
	Are trash cans throughout shop less than ½ full?	
	Are workbenches clean, with the area around them clear? No trip hazards?	
	Are common tools located in or near each bay where they are used (with sufficient quantities of them) such as oil drain pans, jack stands, air tools?	
	Does each bay have retractable air hoses at both ends?	
	Does each bay, where PM is done, have an overhead dispenser for motor oil?	
	Are less frequently used tools stored on shadow boards or in other ways that are easy to get to and easy to inventory?	
	Are fire extinguishers inspected regularly (note dates)? What is the condition of electrical cords? Are there other safety or fire issues?	

Priority	Question	Findings
	FLEET SHOP AUDIT FORM *(cont.)*	
	Are oil storage, waste oil areas and pits, clean with minimal spillage?	
	Are all large tools operable? Are compressors, tire machines, and brake lathes, cleaned and serviced regularly?	
	Are broken parts segregated for inspection by a supervisor or senior mechanic?	
	Check 10 vehicles. Are they clean inside and out, consistent with use?	
	Check tire pressure, alignment, size on duals. How many units are within +5% of rating?	
	Perform PM on a random unit. What did you find?	
	Is their adequate space, storage, and lighting?	
	As you look at the stock room area, is it clean and in order visually?	
	Is there a place for incoming and outgoing rebuilds? Are all rebuildables properly tagged?	
	Are all part storage positions clearly labeled? Check 10 at random. How many are correct?	
	Are all tires tagged with the correct tags? Check 10 random tires for good tags.	

5.3 Job Assignment Dissonance Questionnaire

One major invisible issue among the staff of Maintenance Departments is the difference between what the employee thinks their job is and what the Manager thinks it should be.

Symptoms of this situation include staff working at cross purposes to your goals, staff people spending large amounts of time working on areas that you don't think are important, and complaints that you don't understand the "problems" of their position.

This assignment is recommended when some amount of trust is already in place. There are certainly situations where you may prefer not to use this type of exercise, or may prefer to wait until the timing is better.

The assignment is very valuable if you are willing to look at the results with an open attitude. This exercise can be done in writing, verbally, or you can have an outside party conduct the interviews. It should not be tied in any way to salary or performance reviews.

Employee – Explain that you are conducting a study of the Maintenance department and have them describe their duties, including:
- In a typical week, what percentage of your time is spent doing what activity?
- What are your most important activities?
- What activities do you like best?
- What activities do you like least?
- Add any comments that you think could help this study.

Maintenance Manager – Describe the duties of the employee above, including:
- Define the duties of the above position.
- How much time should be spent on each activity?
- What is the most important activity?
- What is the "mission" of the position?
- If you have a formal job description, include it as a third opinion.

How to Use the Results – Look at the areas where there is a difference of opinion:
- Is the staff person correct in their description of the duties/responsibilities?
- Is there something you can learn?
- Has the job developed over time so that much of the effort no longer supports your current needs?
- Can you re-define the job, using the input from the employee, to better serve your department's needs?

When the review is complete, you will have an excellent start on, or revision to, the job description.

5.4 The Pre-Planning Summary

5.4.1 Projects to Understand the Conditions in your Fleet
1. Standard Operating Procedure Book (SOP):
    ◊ The paper flow diagrams show how information already flows through your organization.

- The vocabulary defines all the standard and special terms used in your organization.
- The new organizational charts show both demands (users) and resources (mechanics).
- The Work Rules section has all of the information about the definition of a workday.

2. Reports:
   - You have a concise reporting hierarchy for reports from your proposed system.
   - This can be derived from the organizational charts.

3. Complete System Sizing and Scope:
   - The paperwork analysis will yield the number of ROs per month.
   - The asset list will show the number of units under maintenance control.
   - The organizational plan will show the number of mechanics.
   - The physical inventory will determine the number of line items that the computerized inventory system will have to store and process.
   - The Asset List and Physical Inventory:
   - This will show the size of the system and can help estimate the effort level to install or maintain a CMMS (Computerized Maintenance Management System).
   - Discussions for additional resources from the organization should always mention the asset base as one of the arguments.

If you are planning to computerize (or re-computerize), these steps will define the Master files for your organization. All of the work so far can directly, or indirectly, be used in the computerization effort. If you are in the purchase cycle for a system, then a summary of these steps will make an excellent system specification.

5.4.2 Introduction of Your Unique Vocabulary

Interestingly, fleets don't use consistent language. In fact, in some cases, different divisions of the same companies don't even speak the same languages (in the maintenance shop). This small project will make it easier for people to communicate within the whole organization. Establish a common vocabulary, first within your department, and then in your organization. Collect the words you use every day.

Include words for:
- Repair Order Unit Number
- Trouble Ticket
- BAF (Bust and Fix, Non-PM items)
- Driver Report 1st Class Mechanic
- Utilization 2nd Class Mechanic
- Repair Reason
- Schedule Miss
- Equipment Class Stock Out
- DIN (Do It Now)
- Recall Scheduled
- Idle Emergency
- Waiting
- Preventive Maintenance
- PM Work Order
- Short Repairs
- Predictive Maintenance
- FIS (Fleet Information System)

Design a poster with these words and hang it where everyone can see it.

The ATA (American Trucking Association) designed a standard system and language for expressing the concepts of Fleet Maintenance. The standard is called the VMRS (Vehicle Maintenance Reporting System). The complete manual is available from the ATA Management Systems Committee whose address is in the appendix.

We will introduce concepts from ATA VMRS throughout this book. In this section we will introduce the vocabulary adopted by VMRS.
- Fleet ID Number
- Work Accomplished
- Utilization Failure Code
- Fleet Code Reason for Repair
- Type Code Scheduled
- Check Digit Non-Scheduled
- Repair Order Emergency

- Activity Code Repair Site
- System Code
- System-Component-Assembly

GENERAL DISTRIBUTION ORGANIZATION	
MAINTENANCE FACILITY WEST STANDARDIZED VOCABULARY	
Assembly Code	Part of the description of what work was done. The assembly defines the specific area worked upon. Example: Air Cleaner assembly or front brake assembly. Consult SOP manual in dispatch for all assembly codes.
DIN Jobs	(Do it Now) Jobs which were not assigned in the morning and which interrupt the job you are on—less than 1 hr. in duration. Duration greater than 1 hr. is called emergency work.
Emergency Jobs	Jobs which were not assigned in the morning and which interrupt the job you are on (over 1 hr. in duration).
Equipment Class	Groups of like equipment in like service. 12 bay straight trucks = class 1200, 16 bay = class 1600, etc.
Fleet ID Number (FID)	Unique number assigned to each truck when it is purchased. FID must be used on all RO, Fuel Tickets and Trouble Tickets.
Failure Code	Why did the part fail (broken, worn through, bent, etc.)? See back of RO for complete list.
Non-Scheduled Work	All emergency and DIN jobs.
Preventive Maintenance Activity (PM)	Inspection, Adjustment, Lubrication and associated repairs up to 1/2 hr. All problems uncovered needing repairs over 1/2 hr. will be written onto ROs.
PM Repairs	Also called Corrective maintenance. Repairs generated by PM inspectors. Scheduled work.
Repair Order (RO)	All work except DIN and Emergency must have a RO issued before the work is started. Get RO from John Kelly. RO required for any parts. DIN, Emergency RO to be written up when job is complete or when parts are required.
Scheduled Work	All jobs that were issued in the morning or jobs issued as a result of PM activity that do not interrupt a job in progress.
System Code	Describes the general area worked upon. Example: Engine or Cab system. Full description of all system codes on back of RO.

5.4.3 Paperwork Study Including Computer Interfaces

The author would like to thank George Gross of J. George Gross Assoc. and Abe Fineman of ICC for this pre-planning tip.

All organizations use paper (or computerized forms) to control or initiate their repair activities. Over the years, these systems have become more and more complex with control documents/systems added and never removed. This is an excellent study for newer Maintenance Managers to undertake to help them learn their operation in detail.

The purposes of this study are:
- Learn all of the paperwork, computer work and system activity.
- Investigate what forms/systems/screens/procedures can be eliminated.
- Determine which systems/forms can be consolidated.
- Streamline queuing waiting time.
- Plug any holes in the control of your operation.

PHASE I: Collection

Time: Choose a time which spans two accounting periods and includes one complete period. Usually six weeks starting just before a new period.

Procedure: Collect copies of all pieces of paper (virtual or real) that pass through your maintenance operation (3 ring binders work well). This includes screen shots of all repair orders, schedules, E-mails, purchase orders, reports, notes, little black book pages, matchbook covers, telephone logs, everything!

It's important to capture the little slips of paper because they usually indicate holes in your control systems. In some cases, we've seen operations that had excellent computerized systems that were ignored. The operation was actually managed from a black book in the foreman's pocket. These smaller documents can be taped or glued to 8 1/2 x 11 loose-leaf sheets (or scanned and added to a virtual notebook).

As you collect these documents, identify where they came from, number of copies produced, any ideas that you have about them, where they go, who signs (digitally approves) them, who uses them, and any other interesting facts about them. These notes form the basis for the next phase.

PHASE II: Analysis of Paperwork Study

Time: After the six week period is over and all forms have been collected.

Procedure: Begin reviewing the documents and your notes. Keep in mind the five goals of the study. On large sheets of paper (17 x 23 or larger), or with the use of a program such as Visio, begin charting the movement of the major forms through your organization. G. Gross uses columns to show people or stops that the form makes and small symbols that are the shape of the form itself. Time is represented by the vertical axis of the paper.

These charts will show you graphically the complexity and route of your major forms. In one major city, it was found that the repair order made 43 stops before work could begin.

This study usually results in reductions of up to 20% to 30% of your paper flow. The key is looking at all the forms and flows at one time.

5.4.4 Create A New Type of Chart to Help Identify Demands and Resources

When a fleet is audited, an important early step is to isolate all users of the fleet's services and all of the resources available to deliver that service.

Procedure to determine the demands on your department:
- On an 11x17 quad ruled tablet (or larger for large organizations) diagram all of the people (Users) that can legitimately request (demand) your resources (these are your demands).
- Use an organizational chart format (for which excellent software is available).
- Include as much detail as possible. Review the last year or two for unusual requests. Don't forget the semi-legitimate requests (the governor's son's Go-Cart or the president's spouse's Mercedes, etc.). There may be department heads or other employees for whom you work to help "grease the wheels" to get your priorities looked after.
- Include drivers, administrators, production, facility maintenance, safety office, security, housekeeping, fire/life support, warehouse, wherever there is equipment to maintain. Try to uncover hidden users for whom you work.
- Include your own department if you maintain your own compressors, doors, lifts, etc.

Procedure to determine the resources in your department to deliver service:
- List all of your mechanics, including their level of skill.
- Include Yard people, Fuelers, Spotters, Inspectors, Car Washers, and Drivers/Gofers.
- Think of hidden resources that you may not be using to their fullest potential such as: Vendors, Drivers, Warehouse People, Stock Clerks, and Plant Maintenance Craftspeople.

BE ALERT! In many cases, you can transfer responsibility and hours to others. For instance...
- While Drivers are fueling, they can also check the oil, and go through a simple check-off list of safety items (part of CDL-Commercial Driver's License already).

- Many vendors have programs of stocking shelves and refilling inventory.
- Stockroom clerks can enter parts used on ROs into computer systems. Many firms have the Stock clerk as one of the main data entry people.

5.4.5 Demand Hours

Demand hours are the hours that the vehicle is in use. For a police fleet, fire trucks, and over-the-road trucks, for example, the demand hours approach 24 hours. For a building inspector vehicle motor pool, the hours might be 7:00 am to 4:00 pm. A municipal subway fleet might have 19 demand hours a day, from 5:00 am until midnight. A school bus fleet might have a split demand, working from 6:30 am to 9:30 am and again from 2:30 pm to 4:30 pm, with a few units out until 6:30 pm. All fleets have characteristic demand hours.

This is important. The bulk of your maintenance activity should be scheduled when your vehicles are not in demand. If you are forced to perform repairs entirely during demand hours then (according to Ron Turley, mentioned elsewhere) your spare vehicle ratio might have to double from 3% – 6% to 6% -12%.

Ideally, the day shift concentrates on multi-day jobs and emergency work. The evening shift focuses on PM and corrective jobs (less than one shift). In some shops, where the demand is during the day and the shop is constrained to day time shop hours, some mechanics come in very early to get some of the quicker jobs out of the way, before the drivers arrive, and others stay late to keep jobs going.

Choices When Your Demand is 24 Hours

If you are a large fleet, staff whenever there is demand. Since it is easier to get parts and outside services during the day, the day shift might be heaviest. This is a tougher situation for a small fleet. In this case, you might resort to an outside service, 24 hour garage, call-out list, or make a deal with another local fleet needing coverage to share the responsibility.

5.4.6 Isolate All Rules That Impact the Workday

The question in this section is how much resource is available to you. This is an information gathering exercise. You may have less (or more) resources than you thought.

Procedure to Determine the Resources Available to You:
- Obtain a copy of your work rules, employee booklet, and/or union contract.
- Consider the full workday and gather all the informal work rules (such as 15-minute wash-up before lunch and end of day, tolerance for personal time during the day, etc.).
- Summarize both formal and informal rules.
- List work restrictions (what class of person can do what type of work).
- Isolate time available for work after considering:
- Breaks
- Wash-up
- Holiday
- Time Off (sick, personal, jury duty, first day of hunting season, etc.)
- Any other restrictions
- Customs such as working in pairs
- Effects of seniority on work assignment

Examine your ability to change rules that have an adverse impact on your operation.

5.4.7   Conduct a Complete Equipment and Vehicle Inventory

The equipment inventory is a two step process. When conducting your inventory, include the following equipment:
- All types of power units, tractors, trucks, cars, fork trucks, pallet trucks
- Chargers and batteries for all electric power units
- Maintenance on building, machinery, etc.
- Trailers, including vans, chassis, flat beds, bulk, tankers, specialized
- Reefer units on trailers, trucks and containers
- Agricultural implements
- Rail Cars, including bi-modal, boxcars, sidings, tankers, rail maintenance
- Construction equipment, including graders, loaders, dozers, compressors, etc.
- Stationary engines, generator sets, co-gen sets

- Overhead bridge cranes
- Ships, container cranes, break-bulk cranes, dockage, piers

On the first pass, create a catalog of the following:
- All power units: Make, Model, Year, License
- All non-power units: Make, Model, Year, License
- All stationary units of all types: Manufacturer, Model, and Year
- All Material Handling units: Manufacturer, Model, Capacity, and Year
- All of your own tools, compressors, doors, buildings that you maintain
- Review repairs for one or two years and include all the "golf cart" and "lawn mower" class units

On the second pass:
- Add detail to line setting ticket level.
- Identify special skills required to support units.
- Insert approximate market value.

5.4.8 Conduct Physical Inventory in the Stockroom

The cost of replacement parts, as a percentage of the fleet budget, has been increasing for the last 25 years from 32% to 44% of the maintenance dollar. Physical inventory is the first step to control usage of, and expenditures on, parts.

Procedure for Physical Inventory:
- List all of the parts (generate stores catalog of everything on the shelf).
- Include qualities and last costs.
- Identify a vehicle group, type, or make/model for each part.
- Segregate any parts that are for vehicles that are no longer part of the Fleet. Investigate ways of converting these parts into cash or usable inventory.
- Segregate large expensive parts, where you have an excessive quantity in relation to your use, that have been in inventory for a number of years. Investigate ways of converting these parts into usable inventory.
- eBay and other auction sites are extremely useful to source and dispose of obsolete parts.

Caution: If the part is unusually difficult to get, or downtime on that equipment is very expensive, keep the part. Consult the section on parts inventory for specific rules about inventory evaluation.

5.4.9 Prepare a Private Budget

The installation of any new system will affect your budgeted expenditures. You will have extra labor until people get used to the new systems and extra parts as you bring your fleet up to maintainability. In this pre-planning step, you will prepare a budget that includes the costs of the new system.

Procedure for Preparing a Budget:
- Copy several years' budgets.
- If the information is available note hours, parts' dollars, and number of units in your fleet.
- Estimate (if the information is not available) emergency hours, scheduled hours, repair incidents, and any major shifts in fleet make-up.
- Add the effects of your new maintenance system (higher PM costs, effect on breakdowns, etc.). This is a completely informal process at this point.

This is your economic impact statement. You can keep it private for now and refer to it quarterly to see how well your predictive powers have been. This information will be used later as a component of a more formal budget process.

5.5 Selling Improvements to Management

In every organization, some issues will be more important than others. Sell your improved maintenance management investments using these issues.

Review the list that follows as if you were top management. What items are the most important? Be sure when you assemble a proposal for top management that you include these items in their language.

Use the same concept for presentations to your operations group. Review the list and pick the two or three top items for operations management. Frame your presentation to the operations group using these concepts.

Important: Use the language (and issues) of your organization to sell Maintenance Management investments.

From the list below, choose the two or three most important issues for each sales pitch:
- Reduce the size and scale of repairs
- Reduce downtime
- Reduce number of repairs
- Increase vehicle useful life
- Increase driver/load safety
- Reduce accidents due to mechanical failures
- Reduce road calls
- Reduce road call associated lateness (time sensitive loads)
- Reduce overtime for responding to emergency breakdown
- Reduce repair time and improve quality due to field repairs
- Reduce use of over the road repair vendors
- Increase vehicle availability
- Reduce spare units required
- Reduce fuel through better maintenance, specification, and operation
- Increase control over parts
- Improve information available for vehicle specification
- Lower overall maintenance costs through better use of labor and material
- Lower cost/mile (KM)
- Improve identification of problem areas in order to focus attention on those areas

Remember to speak about the topics and issues of interest to them! Answer the question, "What is in it for me?"

When you sell a new program to:

Staff – Stress quality of life issues, professionalism, no reduction in crew size, higher job classification, more profitable/stable company

Operators – Stress fewer and less sever breakdowns, improved safety, higher productivity, more profitable/stable company

5.5.1 Charge-Back

One way to get and keep the attention of your user and management groups is to use charge-backs. Each month all of the four cost areas, ownership, maintenance, operating, and overhead, are charged back to the departments using them.

Charging Labor – The labor rate is "burdened" with all associated costs (discussed in depth in the section on Maintenance Costs and the chapter on True Cost of Labor). Typically, the burdened rate is about 2.2 * the Labor Rate, or about $55.00 an hour for a $25.00/hour labor rate.

Charging Parts – Parts prices are marked up to cover all of the costs associated with having inventory (this issue is discussed in depth in the section on Maintenance Costs and the chapter on True Cost of Parts). A $50.00 battery might be marked up 13%, and charged at $57.50.

Example of charge-back:

XYZ INTERNATIONAL CORPORATION
FALLS RIVER FACILITY

Monthly Charge-back Sheet 1-09

Department	Labor	Material	Total
Administration	6575	4050	10625
Distribution	72000	65500	137500
Warehouse	24100	20500	44600
Sales Fleet	14400	15500	29900
Charge-back	117075	105550	222625
Total Fleet Costs	119975	108550	228525
Charges-to-Fleet Budget	2900	3000	5900

By charging costs back to originating departments, you can highlight where fleet resources are being consumed.

Basis For Making Decisions

6.1 Life Cycle Cost

Top management wants to see numbers that show how improvements will positively impact the budget. The key analysis is Life Cycle Cost.

6.1.1 Replacing Units

Some municipalities have a rule of thumb that states they replace a vehicle when they have spent the purchase price in maintenance costs. That works great until a new engine and transmission pushes that vehicle over its budget and someone, whose only looking at the numbers, decides it's time for that unit to go.

In tough times, whatever formula you use to replace vehicles might be thrown out the window to conserve cash. When that is the case, it behooves you to do an excellent job of maintaining the fleet (because you might be married to that unit quite a bit longer then you anticipated).

This is one of the reasons that, in tough times, you have to resist the tendency to cut maintenance to save money. First, it rarely works except in the very short term. Second, it means keeping equipment longer with less maintenance, and both reliability and safety will suffer.

6.1.2 Specifying Vehicles

Organizations might specify vehicles from history, tradition, or based on a trusted vendor or dealer. In fact, the "specification game" should be based on trade-offs using the cost areas of the life cycle cost.

There are five cost areas that contribute to the Life Cycle Cost. The life cycle cost is the total of all of the five cost areas for the life of the unit. In overall financial terms, **the life cycle cost should be the determining factor in vehicle selection.**

You might need to divide Life Cycle Cost by either lifetime mileage or by life in months. Why is this?

The denominator is essential to show the total costs for delivered products (for example, the cost of the vehicle per ton-mile). Cheap, short-lived units will always look better than more expensive, long-lived units without knowing the extra life of the unit (expressed in months, miles, hours, or ton-miles). If the life of the unit is unimportant (such as needing units to cover a three year contract and then discard them) then the cheaper unit might be the best business decision.

Since life cycle costs are cost projections, they are guesses about the future of fuel prices, maintenance costs, and other factors. There are two ways to evaluate life cycle costs, which are different only in the way they handle the time value of money. The first method disregards the time value of money and looks only at the estimates of the total costs. The second method includes the time value of money and weights the investment by when it occurred.

6.1.3   Method One for Determining Life Cycle Cost

Life Cycle Cost (LCC) = ownership costs + operation costs + maintenance costs + allocation of overhead costs + downtime costs

6.1.4   Method Two for Determining Life Cycle Cost

Life Cycle Cost (LCC) / Month = ownership costs + operation costs + maintenance costs + allocation of overhead costs + downtime costs / calculated months of use or ownership

6.1.5   Evaluating Life Cycle Cost

- Utilization (mileage or hours) – High guesses for utilization will skew the results toward low operating cost units. Low utilization guesses will skew away from high fixed cost units.
- Interest rates – High rates will favor lower initial investment (higher operating and maintenance cost) units.

LCC should drive the vehicle replacement decision. When the monthly LCC of an old vehicle exceeds the monthly cost of a new vehicle then the old vehicle should be replaced.

There are four schools of thought on vehicle replacement:
- The "**Run Them** till They Drop" school of fleet operation. In this method the vehicles are retired by rust, multiple major failures, accident, etc. This method requires a lot of attention from the fleet management since equipment is always breaking down

at inconvenient times in inconvenient places. With a "genius" mechanic in charge it can also be a very inexpensive way to run a fleet. Fleets where public image is not an issue can get away with this strategy. You would never catch Coke or Pepsi doing this but you can commonly find your local ferrous scrap company running 'em till they drop.

- The "Wait 'til We Get Tired of Looking at Them" school is a variation of the "Run Them 'til They Drop" school. In this school, the equipment is retired when someone decides that they are tired of seeing the unit around. This usually takes place just after a major expense such as an engine rebuild.

- The "Turn 'em Over" school of fleet management calculates the maximum economic life before major failures and trades the unit while it still has good life. The vehicle is then maintained to insure that it lasts its entire life cycle. A fleet run this way requires less management (far fewer emergencies). Rental car fleets traditionally turn over their cars within the warranty period. They have few maintenance headaches and still get high sale prices.

- The "Compare and Analyze" school compares the cost of operation against a financial model of new vehicle costs. Fleets of this type look at historical costs, add in the costs of downtime and disruptions to the operation, and compare them to the new vehicle model. This school of fleet operation is calculation intensive. It requires periodic recalculation and re-evaluation of fleet condition. Substantial savings are available from the management of the life cycle cost.

6.2 Five Costs of Fleet Operation

Your job, as fleet manager, is to reduce costs and hold the line on the budget. Good management involves reducing and controlling the resources you consume such as fuel, labor, and parts. The overall long-term goal of maintenance management is to control and then reduce costs.

To get a handle on the costs, you must identify the costs at an actionable level, that is, at the level where the consumption of the resource takes place. The resources the fleet consumes are called costs. The costs of a fleet operation can be broken down into five areas. To reduce costs of fleet operation you must reduce costs in one of the five cost areas.

Some costs of fleet operation are difficult to change while others are relatively easy to change. For example, it would be relatively difficult to lay off mechanics and staff to reduce costs. It would be relatively easy to cut fuel consumption through more efficient routes, loads, or (less easily) equipment selection. In the following table, the five cost areas are evaluated in terms of ease of change (ease of improvement).

EASE OF IMPROVEMENT IN FLEET MAINTENANCE COST AREAS	
Area of Cost	Ease of Impact
Ownership	Difficult to Change
Operating	Easier to Change
Maintenance	Moderate Difficulty to Change
Overhead	Very Difficult to Change Substantially
Downtime	Moderate Difficulty to Change

John Dolce, a leading fleet manager and lecturer, suggests the following breakdown of costs and opportunity for improvements (his analysis does not include downtime):

COSTS AND POTENTIAL IMPROVEMENTS BY AREA			
Cost Area	% of Total	Potential Improvement	% of Total
Maintenance Labor Costs	26%	10-25%	2-6%
Maintenance Parts Costs	11%	10-15%	1-3%
Operating Costs	23%	5-10%	1-3%
Ownership Costs	25%	10-20%	0.5-5%
Overhead Costs	15%	1-2%	0.5-1%

Consider all programs to reduce costs and all long-term opportunities for cost reduction in terms of this chart. We stress long-term reductions in the cost structure of your fleet because in maintenance it is easy to affect short-term reductions in exchange for higher long-term costs.

We will now discuss the specific components of the five cost areas.

6.2.1 Maintenance Costs

- True cost of inside labor (includes benefits, fringes, lost time, and overhead)
- True cost of inside parts (includes cost to carry, shrinkage, spoilage, etc.)
- Outside Labor (vendors)
- Outside parts (suppliers)
- Change Oil/Antifreeze

- Road call-accident
- Road call-fuel
- Road call-mechanical breakdown
- Road call-tires
- Misc.
- Hidden costs of failures

Maintenance Costs vary with many factors including utilization, age, type and condition of your fleet, type of service, driver expertise, mechanic expertise, company policy, vehicle specification, quality of fuel/lubes, and/or location/weather. It is very difficult to determine all of the contributors to a particular fleet's maintenance exposure. Managing maintenance means understanding the major contributors to your fleet's maintenance exposure.

Most of today's maintenance costs reflect wear and tear that took place in the past. Therefore, if you reduce or increase your utilization today, your maintenance costs will not be affected until some future date.

In the 72-month life of your typical vehicle, you would expect to pay 2-3 times the purchase price in maintenance costs. As the unit ages, the maintenance costs increase. Maintenance costs increase significantly when a critical wear point is reached (most organizations try to trade before this critical wear point is reached).

Since many factors influence maintenance costs, investments in various areas will have different effects in different fleets. For some organizations, driver training might have a major impact, for other organizations, specification of heavier duty components will reduce costs most efficiently.

You invest funds today to avoid costs in the future. A comprehensive Preventive Maintenance (PM) program will increase your maintenance expenditures in the beginning (the investment). The return may start 12-18 months in the future.

Management of the maintenance component of fleet expenditures is by far the most complex of the five cost areas. You have to track, analyze, and cross reference between like units an average of four repair orders per month per unit (some 288-300 ROs with multiple repairs each per unit for life). The real answer to the exact factors contributing to your maintenance exposure for your operation lies in that mass of data.

Two of the hidden costs that should be included when evaluating different Maintenance modes are the true costs of Downtime (covered in the downtime section) and the hidden costs of emergency repairs (road calls–all types). We use hidden to mean the disruption to your on-going operation caused by emergency calls. This disruption adds to the cost of the interrupted activity and get's charged on the interrupted job's repair order.

6.2.2 Downtime Costs

Downtime calculations are difficult because they don't directly appear on anyone's budget. Frequently, however, downtime costs can be a deciding factor in the decision to invest in a PM or computer system.

Related to downtime is the concept of demand hours. Demand hours are the hours that the equipment is in demand. A one-shift operation that runs equipment 52 weeks per year (less four holidays) has the following demand hours:

Demand Hours = (8 hours/day x 5 days x 52 weeks) - (4 holidays x 8 hours/day) = 2,048

To take advantage of this limited demand, many organizations that operate one shift reduce downtime by using second or third shifts for PM and certain types of repairs. Where this is possible, many of the costs associated with scheduled downtime are eliminated. Downtime due to emergencies will still have an impact.

Reasons for downtime should also be tracked. Much of downtime is not related to maintenance but rather to accidents and other driver controlled situations such as sabotage, running out of fuel, or abuse of the unit. This reason for downtime is usually captured as "repair reason" on the repair order.

6.2.2.1 Downtime Cost Factors
- Idle Operator or Driver salary
- Replacement unit rental costs
- Load replacement cost for ruined product (time sensitive loads like vegetables)
- Late penalties
- Increase in cargo insurance premium, Insurance premium in case of accident
- Loss of early/on-time incentives
- Intangible costs of customer dissatisfaction, hidden costs, other costs

- Revenue loss
- Cost of capital tied up

Use these costs to calculate downtime cost per hour and/or cost per trip and/or cost per field repair incident.

The second factor to add into the downtime equation is the class (defined as like vehicles in like service) of the equipment down. The raw cost of a container crane down is measured in thousands of dollars per hour. A delivery van for pizza has a considerably lower downtime rate. You might have to have several downtime rates if you operate a heterogeneous fleet.

6.2.3   Operating Costs

Over the 72-month average life of a tractor, at today's diesel and tire prices, you will pay over five times the cost to operate the unit as you did to purchase it.

6.2.3.1   Operating Cost Factors
- Fuel
- Fuel taxes (above those paid at the pump)
- Mileage charges on rental/leased units
- Tire consumption (not repairs)
- Add-oil/ Add-Antifreeze/ Add-hydraulic oil
- Misc. Operating Costs

Operating costs vary directly with utilization or use of the vehicle. Reduction in mileage through better routing, reduced idle time, proper tire inflation, and preventive maintenance that corrects bearing/drive train/clutch problems, will all favorably impact operating costs. The unique aspect of operating costs is that they respond immediately to improvements in your operation.

6.2.3.2   Fuel Costs

Fuel cost is the single largest cost of fleet operation. Fuel costs can be reduced by proper specification of new fuel-efficient engine/drive train combinations.

Ownership costs may increase in order to reduce operating costs. Super light rigs may decrease operating costs while increasing maintenance and ownership costs. A trade-off in the other direction might include heavy duty parts such as axles which will slightly increase fuel/tire costs due to increased weight and may reduce maintenance costs.

As a rule, operating costs gradually increase as the unit ages. In times of rapidly increasing fuel prices, less fuel efficient units dramatically increase in operating costs.

FUEL SAVINGS IN A CLASS 8 DIESEL (100,000 MILES/YEAR)	
4.5 MPG rig	5.5 MPG rig
miles per year / miles per gallon = gallons used per year	
100,000 /4.5 = 22,222 gallons	100,000 /5.5 = 18,182 gallons
22,222 - 18,182 = 4,040 yearly gallon savings	

DOLLAR SAVINGS PER YEAR ON 4,040 GALLONS	
Cost per Gallon	Savings per Year
$1/gallon	$4,040
$2/gallon	$8,080
$3/gallon	$12,121
$4/gallon	$16,161
$5/gallon	$20,202

To calculate total fuel cost for life:

Life Fuel Cost = (miles per year x years x average fuel cost) / MPG

6.2.4 Ownership Costs

Ownership is the first cost of fleet operation. These costs start to accumulate before the first mile is driven.

6.2.4.1 Ownership Cost Factors

- Purchase costs, depreciation, and costs of money
- Lease/rental payments (fixed portion)
- Insurance costs, self-insurance reserves
- Permit, license costs, statutory costs (costs mandated by laws)
- Make-ready costs, new/used vehicle preparation
- Actual cost of searching, shopping for vehicles
- Re-build/Re-manufacture costs
- Labor to strip parts off retiring units

Ownership costs vary directly with the number of vehicles in your fleet. A reduction in fleet size through increased availability (less reserve units), decreased mileage (fewer vehicles required due to more effective routing), or reduced fleet size (through peak demand leases), will all favorably impact ownership costs. Ownership costs must be paid before the first mile is driven.

Ownership costs can be increased by specification. If you specify a 12000GVW axle as opposed to a 10000GVW axle, your ownership costs will increase. Specification of extra trim, options, and radial tires will increase your ownership costs. Note that many of these expenditures, while they may be desirable (by lowering other costs or improving driver comfort/morale), will increase ownership costs. Detailed discussions of some of the costs follow.

6.2.4.2 Depreciation

Depreciation is one of the components of ownership costs that goes down as the vehicle ages. There is an on-going discussion about the best method of depreciation to use for maintenance purposes. Your accounting department uses techniques dictated by the IRS to influence (increase/decrease) profits.

Our goal is to simply consume the value of the unit over its true productive life. We recommend that you be aware of the technique used by your accounting department to depreciate your fleet, however, for maintenance management purposes, use the straight-line method.

6.2.4.3 Depreciation Formulas

- *Total Depreciation = total purchase price - salvage value*
- *Yearly Depreciation = (total purchase price - salvage value) / years of life*
- *Monthly Depreciation = (total purchase price - salvage value) / months of life*

Where:
- Total Purchase Price is the total of all costs, including closing costs and vehicle preparation.
- Salvage Value is a unit's market value at retirement.

- Years of Life is a unit's average age in years at retirement.
- Months of Life is a unit's average age in months at retirement.

DEPRECIATION CALCULATIONS ON A NEW BRAND X TRACTOR	
Total Purchase Price	$82,500
Age at Retirement	72 months or 6 years (on average)
Salvage Value	$7,500
Total Depreciation	$82,500 − $7500 = $75,000
Yearly Depreciation	($82,500 − $7,500) / 6 years = $12,500
Monthly Depreciation	($82,500 − $7,500) / 72 months = $1,041.67

6.2.4.4 Interest

If you took out a loan of $82,500 for 5 years (60 months) at 15% simple interest and no down payment, your monthly payment would be $2406.25 per month.

SIMPLE INTEREST CALCULATION ON A LOAN OF $82,500			
Term		Loan Amount	Interest Rate
5 years	60 months	$82,500	15%
loan amount / term in months = Monthly Principal Payment			
$82,500/60		$1,375	Monthly Principal Payment
loan amount x interest rate x term in years / term in months = Monthly Interest Payment			
($82,500 x 0.15 x 5)/60		$515.63	Monthly Interest Payment
monthly principal payment + monthly interest payment = Total Monthly Payment			
$1,375.00 + $515.63		$1,890.62	Total Monthly Payment

6.2.5 Overhead Costs

Overhead costs tend to be fixed except for major changes in the fleet's size or role. These cost areas tend to be the last to change and are generally only changed after a careful management decision process. For example, you can change Operational Costs by changing to radials, or smaller vehicles, thus reducing fuel costs. However, overhead costs only change as the result of closing facilities, consolidation, outsourcing, re-staffing, etc.

6.2.5.1 Overhead Cost Factors
- Cost of Maintenance Facilities
- Heat, light, power, phone

- All persons in maintenance department not reported on repair orders
- Supplies not charged to repair orders
- Tools and tool replacement
- Repair of Maintenance Facility, maintenance tools
- Clean-up
- Computer systems, all expenses

Excellent maintenance controls may be able to reduce facilities, supply usage, non-repair time, and tool loss. Applying sound maintenance procedures to your buildings and tools will certainly reduce your overhead costs.

Overhead costs begin to accrue before the first truck is purchased and before the first mile is driven. This category of costs is one of the opportunities for savings that the full service leasing agent looks to in saving smaller client money.

6.3 Budgeting for Maintenance

Where is the money invested into the fleet spent? The budget should reflect the five cost areas already mentioned. If it does, the performance to budget becomes the score card to measure the effectiveness of the maintenance program (and the effectiveness of any improvements).

The fleet budget corresponds (usually on a line by line basis) to the cost categories in the general ledger. The categories are not always set up in an intelligent way for fleet analysis so there has to be a cross reference to the five cost areas.

Also, it is sometimes tough to define the downtime costs as general ledger transactions (since the loss might be good will or a calculation based on costs and lost revenue).

| MATERIALS OF ALL TYPES ||
General Ledger Account	Cost Area Account
Parts	Maintenance
Stockroom personnel, purchasing	Overhead
Large spares (engines, etc.)	• Maintenance • Ownership
Maintenance materials	Maintenance
Supplies and consumables	Maintenance
Fuel	Operating

MATERIALS OF ALL TYPES (cont.)

General Ledger Account	Cost Area Account
Fuel tax	Operating
Tires	Operating
Motor oil	• Operating for add oil • Maintenance for change oil
PPE (disposable)	Maintenance
Inside fabrication	• Ownership • Maintenance
Outside fabrication	• Ownership • Maintenance
Rebuilding and remanufacturing services	• Ownership • Maintenance

LABOR

General Ledger Account	Cost Area Account
Inside Maintenance labor – straight time	• Maintenance • Operating – tire changes • Ownership – mounting or modifying a body or make ready
Inside Maintenance labor – overtime	• Maintenance • Operating – tire changes • Ownership – mounting or modifying a body or make ready
Supervisor, manager, clerks, data entry	Overhead
Borrowed labor from other departments	Maintenance
Benefits, workman's comp, FICA, etc.	Maintenance
All kinds of leave	Maintenance
Training	Overhead

CONTRACTS AND OUTSIDE LABOR

General Ledger Account	Cost Area Account
Road call vendors	Maintenance
Outside Labor	Maintenance
Service contracts (shop equipment, HVAC, tires)	Overhead
Vendor work (dealers)	Maintenance
Inspection services	Maintenance
Other Consultants	Overhead

VEHICLES

General Ledger Account	Cost Area Account
Depreciation	Ownership
Insurance	Ownership

VEHICLES (cont.)

General Ledger Account	Cost Area Account
Licenses and permits	Ownership
Make ready and disposal costs	Ownership
Repair truck (road calls)	Maintenance
Pick-up truck	Maintenance
Leases	Ownership
Rentals	Ownership

RENTALS

General Ledger Account	Cost Area Account
Equipment rental	Maintenance
Tool rental	Maintenance
Vehicle rental	• Ownership • Maintenance – service truck
Demurrage (welding tanks, rail cars, etc.)	Maintenance

OVERHEAD AND OTHER COSTS

General Ledger Account	Cost Area Account
Heat, light, power	Overhead
Telephone, computer lines	Overhead
Security	Overhead
Housekeeping services	Overhead
Waste removal (both hazardous and non-hazardous)	Overhead
IT support	Overhead
Damage Repairs	Overhead
Insurance	Overhead

6.4 Macro Budgeting of Labor

This section introduces a technique for yearly labor budgeting. This process can be considered a hybrid of planning and budgeting. Before the technique is introduced, here are some distinctions in the language:

- **Budgeting** – Of the three processes we will discuss, the Budget is the process that is furthest from the actual work. This is a macro view of your operations. If you were budgeted 35,000 hours last year, and your fleet is the same size this year, then you will get 35,000 hours this year. Budgeting is an accounting function. Budgeting requires only a limited knowledge of the maintenance function.

- **Planning** – This is a more detailed view of maintenance. Planning requires knowing the resources required for the various activities. Planning attempts to look at the work at hand and match resources to demands. Planning is a management function.
- **Scheduling** – The Schedule is the most detailed. Scheduling matches the person, tools, materials, location, and vehicle to a time frame. A high degree of knowledge is required to avoid collisions (two people wanting particular tools at the same time) or waiting (vehicle, bay, materials, person not available). Scheduling is a supervisory function, and to be successful, requires constant contact with the shop floor. Scheduling will be covered in the next section.

There are eight gross demands on the Fleet Maintenance Resource. Each of these demands has a magnitude, duration, frequency and time.

These eight demands are:

- **PM Hours** – Based on your fleet size, utilization, and the standard times of PM activities, you can predict how much time PM's will take. Since you have some flexibility in scheduling, you can consider PM's as a level demand.
- **Scheduled Repairs Hours** – As your PM inspectors inspect each piece of equipment, they write-up repairs. Your drivers and operators write-up problems. These write-ups become scheduled, and the repairs are considered scheduled repairs as long as they don't interrupt jobs in process. You can look at previous years to get an idea of the hours for this activity. Since you have control of the schedule, this demand can be considered level throughout the year.
- **Emergency Hours and DIN Hours** – At the beginning of the year, budget the same amount of hours for emergencies as the previous year. At the end of the year, you can back off emergency hours as the PM system starts to take effect. For purposes of budgeting emergencies, create a level demand. In fact, emergencies will tend to bunch. In larger fleets, emergencies will look more and more level. See seasonal demands.
- **Rebuild Hours** – At some point, units that have not been maintained correctly for a period of time, or have reached the end of their useful life in this cycle, will have to be rebuilt. The rebuilding effort should be added to your maintenance budget (separate from any current maintenance activity). If you are doing the rebuilds to bring units up to PM standards then the hours will have to be budgeted. Since you have control of the rebuild schedule, you can use rebuilds as a crew-balancing tool.

- **Seasonal Demands** – Outdoor conditions can be tough on motive equipment. Certain maintenance activity is directly related to the seasons. Review of cooling systems before summer and winter, or recharging air conditioning before summer, are seasonal demands. You can also use this category to pick-up some percentage of the seasonally driven emergencies or seasonally driven PMs. Budget hours at the beginning of each season based on history.

- **Social Demands** – While your primary mission is maintenance of the fleet, you may be called upon for other duties in your organization. These duties are based on your individual history and may include supplying drivers, tours, picnics, or work on non-organization equipment. Estimate your hours for these activities.

- **Expansion** – Any expansion in the size of your physical facility, the size of your fleet, size of your work force, or additions to the scope of your control, will add hours to your requirements. New vehicles require make-ready time. New facilities require disruption to current activities, as well as direct time. Adding satellite facilities will result in additional lost time until systems are well in place. Estimate additional time if an expansion is contemplated.

- **Catastrophes** – Add time for two catastrophes. You can review your records for the actual amount of time spent in a typical catastrophe. This can include floods, hurricanes, trucks taking out the sides of buildings, fires, etc.

Distribute your labor budget to the user group, staff, and top management for comments. Note that you might crew for 80% of demand and use outside vendors during peak periods. If your current available hours are a small percentage of your demand then something will have to be negotiated. You will have to fight for the resources to do your whole job.

6.5 Evaluating Competing Investments

This section is concerned with techniques to evaluate competing investments. Few organizations use all the methods. Determine what techniques are used in your organization, learn them, and use them.

6.5.1 Return on Investment (ROI)

Return on investment (ROI) is the most commonly used measure for investments. ROI is expressed as a percentage of return you earn per year.

Return on Investment = yearly income/total investment

(If the yearly income varies, you can evaluate each year separately, average the years together, or weight the early years higher using a weighted average.)

6.5.1.1 ROI of Common Investments:
- Savings Account 1-2%
- Money Market 2-3%
- Mutual Fund 10% (When things are going well)
- Small Corporate Investment 50%
- Capital Improvements 33%

6.5.2 Payback Method

The second most common method of evaluating investments is to determine the number of years (or months) it will take to pay off your investment based on the investment's return. The payback method is frequently used alone or with ROI to which it is closely related.

Payback in Years = total investment/yearly income from investment

6.5.3 Other Methods of Evaluating Investments

There are many other ways of evaluating competing investments. Some methods are used to pinpoint certain aspects of the investment (such as cash flow analysis or first year performance). There is also a group of evaluation techniques known as present value/present worth that are by far the most complete (and complex) because they take into account when investments occur and when income is received.

6.5.3.1 Cash Flow Analysis

Most organizations use cash flow analysis to plan their overall investment program. A few will look at the cash flow from an individual investment. The idea is to plot the movement of cash into and out of an investment. Some investments (installing a PM system, for instance) require constant monthly outflows of cash for a long period of time (1-2 years) before providing returns. Other investments (such as re-powering a tractor to improve efficiency) require significant cash on day one and provide immediate return.

Cash flow analysis is a powerful tool to coordinate several investment projects. You can alter the timing between projects to minimize the cash out and maximize the overall investment.

6.5.3.2 First Year Performance

This method looks at the ROI for the first year only for the competing investments. If two investments have similar average returns, the one with better first year performance may be the better overall investment. Some organizations that stress cash management (or are short on cash) look very hard at how the investment will act in its early stages.

6.5.3.3 Average Rate of Return (ARR)

This is the same as Return on Investment (ROI) over the entire life of an investment. The ROI will vary from year to year. Once you add all the return and all of the investments together, you can determine the ARR.

AVERAGE RATE OF RETURN	
ARR on a small public warehouse of 50,000 sq. ft with material handling equipment.	
ARR = average yearly income after tax/total investment over life	
Average Yearly Income (after expenses and taxes)	Total Investment (paid entirely with internal funds)
$210,000	$1,400,000
ARR = $210,000/$1,400,000 = 15%	

It's interesting to note that by financing, you can sometimes significantly improve the ARR (or ROI) because the funds can be provided at a lower cost than the return.

6.5.4 Present Value – Present Worth

The value of all money decreases over time starting immediately. Receiving $50,000 today has more value than receiving the same money a year from now, or a decade from now.

Most organizations have internal standards for the value of money over time. The money might be invested in a low risk investment and earn a return. Other organizations set the value of the money at the opportunity cost, which is the return they could receive from those funds invested into another project.

In the late 1970's, the prime interest rate was 20%. This meant that competing internal investments had to provide very high returns to compete with market type investments. During that period, comparatively few maintenance investments could be made. During the late 1990s into the 2000s, the interest rate dropped to the 5-7% level and then to the 4-5% level, which allowed for maintenance investments. The very low subsequent rates make maintenance investments even more attractive.

$$\text{Net Present Value (NPV)} = \sum \frac{R_t}{(1+K)^t} \text{ for years 0 to n}$$

where:
- R_t is the net return (cash flow) for that year
- K is the internal rate of return
- t is the number of the year

Notice that the denominator gets smaller quickly when K is larger. That means if K is larger, then that investment requires a quicker payback to be justifiable.

NET PRESENT VALUE

NPV of an investment in a PM system with either a 10% or 20% rate of return over a five year time period.

$$\frac{R_t}{(1+K)^t}$$

Year t	Net Return R_t	K=10% $(1+0.10)^t$	K=20% $(1+0.20)^t$
t=0	[$50,000]	$\frac{[50,000]}{1.1^0} = [50,000]$	$\frac{[50,000]}{1.2^0} = [50,000]$
t=1	[$30,000]	$\frac{[30,000]}{1.1^1} = [27,273]$	$\frac{[30,000]}{1.2^1} = [25,000]$
t=2	$15,000	$\frac{15,000}{1.1^2} = 12,397$	$\frac{15,000}{1.2^2} = 10,417$
t=3	$25,000	$\frac{25,000}{1.1^3} = 18,783$	$\frac{25,000}{1.2^3} = 14,468$
t=4	$45,000	$\frac{45,000}{1.1^4} = 30,736$	$\frac{45,000}{1.2^4} = 21,701$
t=5	$45,000	$\frac{45,000}{1.1^5} = 27,941$	$\frac{45,000}{1.2^5} = 18,084$
∑ (sum)	$50,000	$12,584	[$10,330]

Factoring in the time value of money, after five years, the $80,000 investment in a PM system at a 10% rate of return generates a positive net present value ($12,584). However, at a 20% rate of return it is negative (-$10,330).

6.6 Effect of Maintenance Investment on the Bottom Line

You are in an extremely competitive battle for the organization's investment dollars. Investments in maintenance can earn big returns. You must sell to your strong suit, which is cost avoidance, improved customer satisfaction, and reduced downtime.

In the past, Maintenance Departments have not done as good a job promoting their importance to the bottom line in order to secure company resources as have other departments in the organization. In fact, Maintenance, in some organizations, is the worst at selling real return on investment, and, more importantly, the effect of maintenance investments on the bottom line.

Maintenance investments flow directly to the bottom line. In non-profit fleets, the language is different but the effect is the same. Same students transported for lower dollars, same streets plowed for fewer dollars.

It's much easier to sell a half million dollars in sales than $25,000 to $50,000 in maintenance savings (especially if the proof takes more than three sentences). Additionally, if the sales department has also provided a professional presentation, and invited the president to a meeting with the prospective customers, management is already involved. This just makes your job tougher, not impossible!

Maintenance Costs 7

The bulk of the effort in fleet management is to control maintenance costs. These costs are complicated and changing them (and the activity they represent) can have severe consequences.

7.1 Longer Life with Less Effort

The goal sought by fleet users is longer life with no investment in maintenance. The unit should just run smoothly every day. Of course, much of the profit of the OEM's comes from their parts and (for some of them) their service business. The OEM's interests and the user's interests are somewhat different.

You need to look closely at what works (components that operate without too much maintenance) and improve, change, or redesign, what doesn't. Maintenance Improvement is simply longer life with less effort.

The curves in the following illustration represent the average life (or MTBF – mean time between failures) of components in a fleet of like vehicles in like service, such as engines in class 8 tractors in long haul configuration (or bearings, seals, etc). Each curve is a different maintenance strategy. Three strategies (sometimes called scenarios) are represented: Do Nothing (called Bust 'n' Fix), PM, and Maintenance Improvement.

Think about a bag of popcorn with each kernel representing an engine. When you heat the popcorn, some kernels pop early (premature failures – infantile mortality), some pop late, but the bulk pop in the middle of the process. If you had a large enough population of kernels, and could measure the time it took each to pop, the curve would start to approach a bell shaped curve.

When the kernels pop is represented by the distance to the right of the Y-axis (distance along the X-axis which is equal to time) and how many pop is represented by how high the curve rises (Y-axis is equal to the number of events). The actual MTBF number will be influenced by how the components are used, how well they are engineered and built, how well they have been maintained and the conditions under which they are operated.

Figure 1: Longer Life with Less Effort

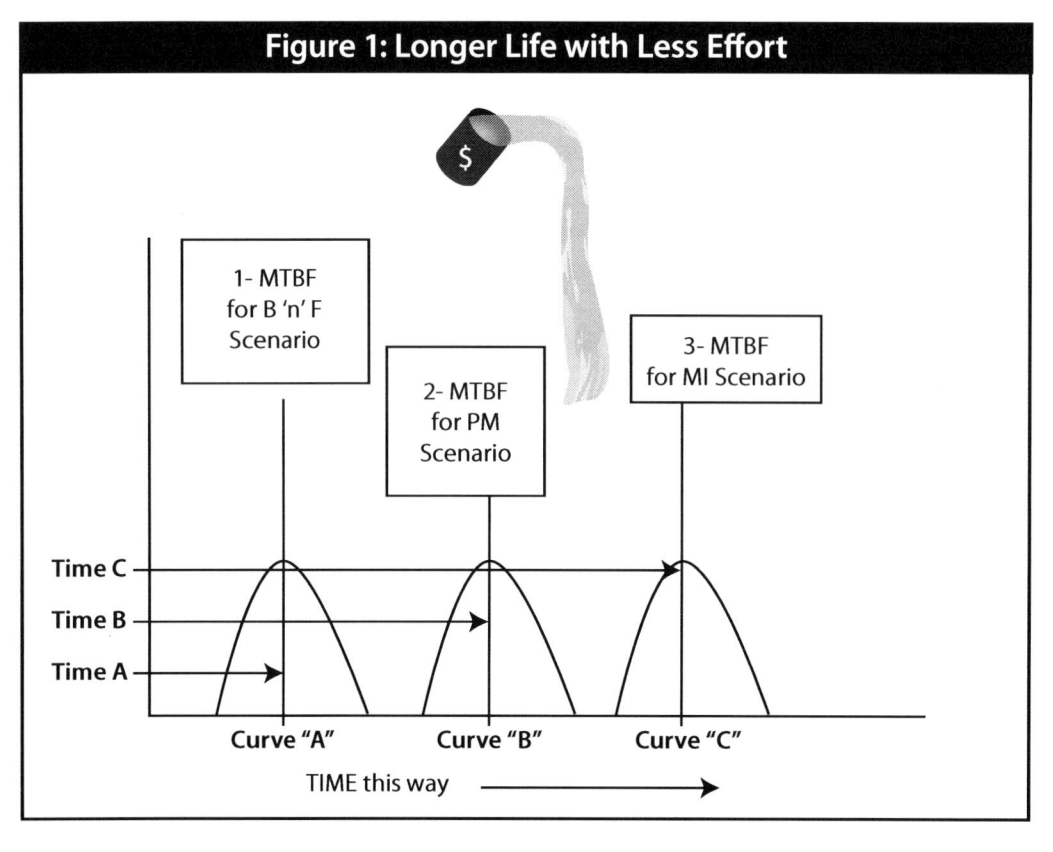

- 1- MTBF for B 'n' F Scenario
- 2- MTBF for PM Scenario
- 3- MTBF for MI Scenario

Time C, Time B, Time A

Curve "A"   Curve "B"   Curve "C"

TIME this way →

Curve "A" shows the "Bust-n-Fix" scenario, where assets are allowed to break down naturally. Left alone, with no maintenance of any kind until failure, each cylinder, engine, or other component will deteriorate and fail in a given amount of time (represented by curve A, at the extreme left). The mean of this curve is Time A. With greater numbers of cylinders, the graph will look more and more like a normal distribution or bell shaped curve (our diagram is simplified because differing failure modes will make the actual distribution more complex).

Curve "B" represents improved life resulting from PM and predictive maintenance. You are cleaning the cylinders, or other components, keeping everything tight, adding an appropriate amount of lubrication, and so on. Notice that the life curve has shifted significantly to the right showing longer average life. The mean of this curve is Time B. To stay at this level, funds have to be continuously invested in the form of PM labor and materials. As soon as that flow of money stops, the curve will slide back to "A".

Herein lies the worrisome aspect of PM. When a new manager comes into a facility, there is the temptation to want to look good and shine. If the manager chooses to cut back on PM as a means to that end (to reduce costs), the chart shows that there will be no impact on failure rate (and therefore reliability) until enough time passes for the curve to decay to "A". That time lag could be a year or more. This is known as "looting the operation".

The profit will go up in the short term (costs will go down). The new manager will be perceived as a hero for increasing profit. However, this is only until Curve B reverts to A. If the manager leaves before the curve decays, he or she will leave a hero and the next manager is stuck with the results of the bad decisions.

Curve "C" is the goal of maintenance. What if you could impact the failure rate in a more permanent way? What if you could make progress and change the nature of failure in your fleet forever? This is called maintenance improvement, where the life of the unit unattended is longer than either "A" or "B". The mean time between failures of this curve is Time C. This improvement can be the result of using better seal kits, better oil, or higher quality components. The key is increased MTBF without having to pour money in every month.

This improved ratio is the ultimate goal in maintenance, so this is where your attention should be focused. However, the maintenance improvement should be logical. You would not want to spend $50,000 to avoid a $100 problem. All improvement efforts in the maintenance department should have permanent maintenance improvement as a goal.

Failure analysis is a major aspect of longer life with less effort. By identification of problems that repeat, and uncovering the why of the failure, you can get more mileage with the same maintenance effort.

7.2 Unfunded Maintenance Liability

PM systems fail because PRIOR SHORT-SIGHTED DECISIONS wreak havoc on any fleet manager trying to change from a fire fighting (B'n'F) operation to a PM operation. Even after running for several months, there are still so many emergencies that you may feel as though you will never make headway.

You face unfunded maintenance liabilities. The only way through is to pay up, modernize, and rebuild yourself out of the current situation. This is where the investment must be made. Any sale of a PM system to top management must include a non-maintenance budget line item for past decisions.

Remember, the wealth was removed from the equipment without maintenance funds being invested to keep it in top operating condition.

7.3 True Cost of Labor

All maintenance decisions are based, in part, on the number of hours and dollars needed to maintain or repair a unit. The number of dollars should be based on the charge rate of the mechanic, not just their salary.

You pay a maintenance worker a certain number of dollars each month. Yet it costs even more to have him or her on the payroll. The true cost of labor includes all the costs to employ someone. All calculations should be based on the true cost of an hour of maintenance labor.

The true cost of labor includes:
- Direct wages
- Overtime (percentage factor)
- Benefits
- Indirect costs

These costs must be spread over the time actually spent on chargeable repairs.

Benefits include the costs of health insurance, FICA (Social Security—employer's contribution), pensions, life/disability insurance, and any paid perks.

Indirect costs include indirect salaries (all support people who don't show up on ROs), materials (ones not charged on ROs, such as bolts, grease, etc.), supplies (uniforms, bulk materials, soap, etc.), costs of the shop/yard (utilities, depreciation of facility and tools, insurance, taxes), allocation of costs of corporate support, costs of money, and hidden or other indirect costs.

The following exercise is exactly the same one that a maintenance vendor would use to decide what to charge you as a shop rate. The only difference is that their calculation includes a profit.

Cost per Hour Calculation:
- Direct hourly wage: $25.00
- Benefits @ 25% of wage: $6.25
- Indirect costs: $9.00

Total Cost per Hour: $40.25

The total cost per hour must be increased to cover time paid that doesn't appear on ROs. Consider the 2080 hours straight time (52 weeks x 40 hr/week) available per year.

Annual Time: 2080 Hours

- Vacation – 160 hours – 4 weeks
- Holiday – 64 hours – 8 paid holidays
- Paid Sick Leave – 40 hours – 5 sick days
- Other – 32 hours – 4 days per year for all other categories including jury duty, National Guard, training, union, etc.

Total Available Annual Time: 1784 Hours

Ratio (annual hours to available hours):

$$\frac{2080 \text{ (annual hours)}}{1784 \text{ (available hours)}} = 1.166$$

Actual Cost per Hour Calculation:

- Total cost per hour (from above): $40.25
- Ratio (from above): 1.166

$40.25 *(total cost per hour)* x 1.166 *(ratio)* = $46.93

Actual Cost per Hour: $46.93

If all of the available hours were charged on ROs then the true cost of labor would be $46.93. However, even in a closely controlled shop, some time is left uncharged. Add a 10% factor for time working on uncharged jobs (sweeping up, small repairs that are not logged, etc.)

True Cost per Direct Labor Hour

- Actual cost per hour (from above): $46.9

$$\frac{\$46.93 \text{ (actual cost of time)}}{0.90 \text{ (10\% uncharged time)}} = \$52.14$$

True Cost per Hour: $52.14

True Cost per Minute: $0.87 (rounded to $0.90)

That's $0.90 every minute when looking for tools, waiting for the parts attendant, or finding the manual for the computer codes, etc.

TRUE COST OF LABOR WORKSHEET

Location: Main Garage
Grade: Class 1 Heavy Duty Equipment Mechanic

COST OF DIRECT WAGES

Wage Rate	Amount
Base Wage	$30.00 $45.00 OT Rate
Cost of Benefits = $30 *(base rate)* x 22.5%	$6.75
Total Wage = $6.75 *(cost of benefits)* + $30 *(base wage)*	$36.75

INDIRECT COSTS ALLOCATED BY REPAIR ORDER HOUR

Yearly Indirect Charges	Amount
Indirect Salaries	$150,500
Bulk Materials	$37,200
Supplies	$31,500
Cost of Shop and Yard	$62,900
Depreciation on Tools	$29,700
Misc. Costs	$21,500
Total	$333,000
Estimated Total RO Hours: 29,050	

INDIRECT COST CALCULATION

Formula	Amount
Indirect Cost/Hour = $333,000 *(total indirect costs)*/29,050 *(estimate of total RO hours)*	$11.46
Total Wage = $36.75 *(total wage from above)* + $11.46 *(indirect cost/hour)*	$48.21

INDIRECT COSTS OF PAID BUT UNAVAILABLE HOURS

Type of Hour	Paid Hours
Vacation	120
Sick Leave	80
Holiday	80
Training	32
Other (Union, Jury, etc.)	32
Total	344

Formula	Amount
Total Available Hours = 2080 *(total annual hours)* − 344 *(total unavailable hours)*	1736
Ratio = 2080 *(total annual hours)* / 1736 *(total available hours)*	1.192
Total Wage = $48.21 *(total wage from above)* x 1.192 *(ratio)*	$57.47

COST OF UNCHARGED TIME *(cont.)*	
Formula	Amount
Total Wage = $57.47 *(total wage from above)* / 0.90 *(10% uncharged time)*	$63.86
*$1.10 per minute (round up from $1.06)	

7.4 True Cost of a Part

Consider the distinction between the price of an inventory item and the cost of that item of inventory. This is the same argument we pursued with the true cost of labor. There are certain costs associated with having parts on hand. These costs have to be spread over all the parts used and added to the part price to yield the part cost. The ratio is called the charge out rate.

If you ran a parts business, you would lose money if you sold the parts for what you paid. If you gradually increased the selling price, at some point you would break even. The breakeven price is the one you seek for a non-profit parts storeroom (you want to recover costs but not make a profit from your internal customers).

These costs, which are unique for each operation and can be unique for each location within an organization, include:

- **Cost of Money:** Your organization could invest the money tied up in inventory into other business opportunities, or at market rates, and get a secure yield from 5% to 7% in today's market. This cost of money is often called the opportunity cost.
- **Expenses of Warehousing:** This includes depreciation on building space and shelves, an allocation for utilities, building maintenance and security, life cycle costs on material handling equipment, forms and paper, office supplies and machinery. It is simplified if you rent the building. Depending on the lease terms, your expense of warehousing is simply the rent. Cost is usually figured at 2.5% to 6.5%.
- **Taxes and Insurance:** Some localities tax the assets of the organization, most have real estate taxes, and this includes casualty insurance. Even if you self insure, there is some exposure and some cost. Costs vary from 1% to 3%.
- **People Costs:** Includes full-time and part-time stock clerks, allocation of other clerks, pick-up/delivery people, supervisors, and purchasing agents. Overall Parts volume has a large influence on this cost. It varies from 14% to 40%.

- **Deterioration, Shrinkage, Obsolescence, and Cost of Returns:** Includes the cost from parts that become unusable, damaged, disappear, are obsolete, or incur a re-stocking fee. Costs vary from 4% to 15% (higher if shrinkage is a significant problem).

In *Fleet Management* (McGraw-Hill, 1984), an older but excellent fleet text, the author, John Dolce, states, "It takes 12% – 15% of a company's annual parts expenditures to support people costs...Support dollars... should be approximately 25% of the on the shelf inventory."

From Dolce's estimates, there are two components to the overhead of carrying inventory. One component is based on a 12% or 15% cost for all the acquisitions for the year (sometimes called the COA – Cost of Acquisition). The other cost is the 25% COO – Cost of Ownership.

The chart below combines the two costs for 12% COA and 15% COA. If you assume that you can turn your $60,000 inventory four times per year, then the costs are as follows.

CALCULATION FOR PARTS CHARGE-OUT RATIOS CAPTURING TOTAL PARTS RELATED COSTS

Inventory (turns 4 times per year)	$60,000
Annual Parts Expenditure	$240,000

Costs		
Percent	Calculation	Result
COA	(Inventory Level x COO) + (Purchases x COA)	Carry Cost
12%	($60,000 x 25%) + ($240,000 x 12%)	$43,800
15%	($60,000 x 25%) + ($240,000 x 15%)	$51,000

Cost Ratio for Inventory Charge-Out		
For Percent	Calculation	Multiplier for Parts Charge
12%	1 + $43,800 / $240,000	1.183
15%	1 + $51,000 / $240,000	1.213

Using 12% COA, a $100 battery would be charged at $118.30 (at 15% COA, the up-charge would be $121.30) on the Repair Order. This up-charge would fairly capture the costs of inventory and purchasing.

7.5 Breakdowns

In your presentation to management, consider the important benefits specific to reduced breakdowns. Decide which benefits would be of greatest use to your organization and feature them at the beginning of your presentation.

Breakdowns are one area where fleet maintenance effectiveness directly affects departments outside the fleet domain. Breakdowns affect the sales and marketing staff, as well as, operations. This can be a strong selling point if you can develop data on the costs. Be sure to point out that the parts and labor costs of a breakdown are like the tip of an iceberg that you see, but the rest of the costs are hidden and will end up on someone else's budget!

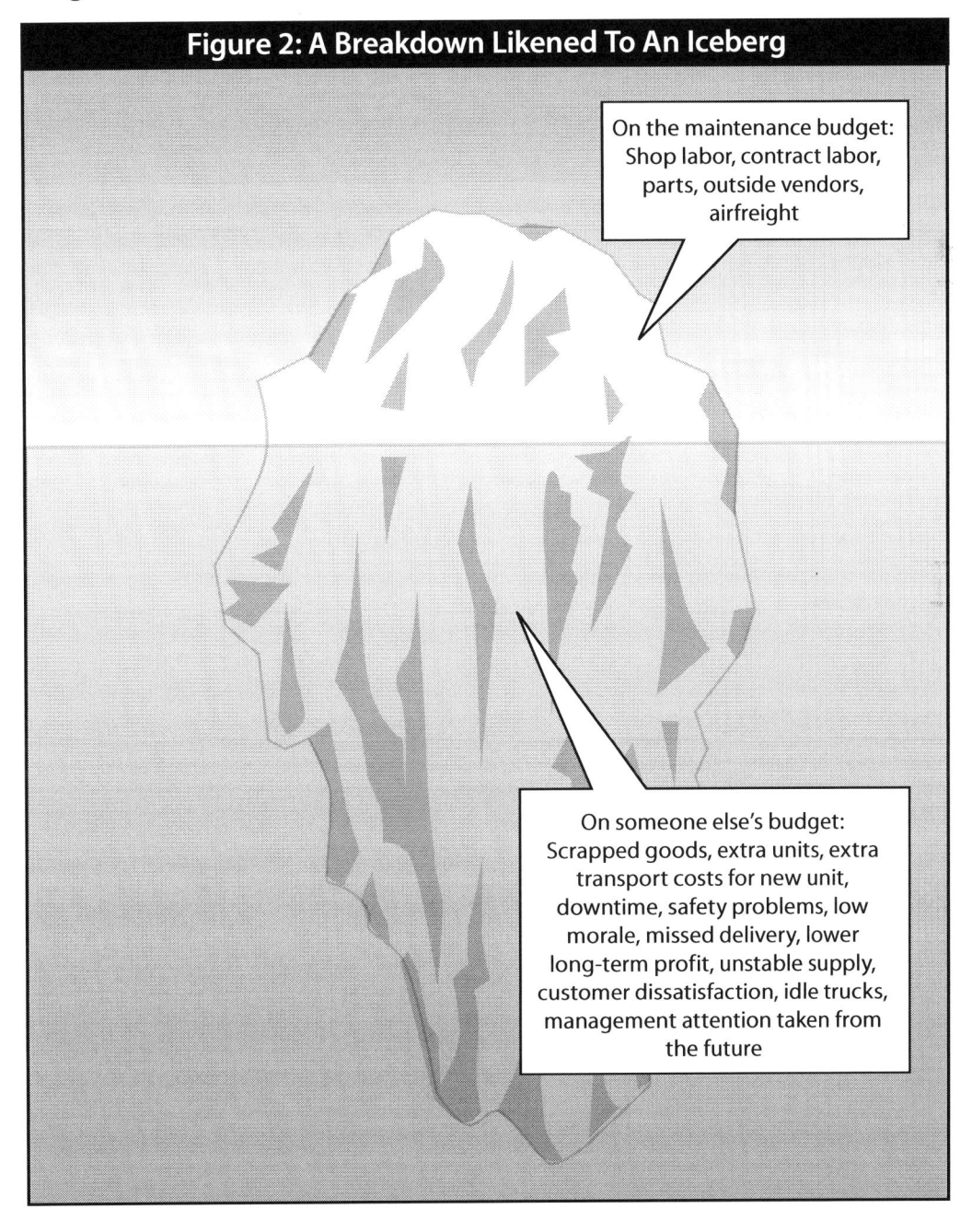

Figure 2: A Breakdown Likened To An Iceberg

On the maintenance budget: Shop labor, contract labor, parts, outside vendors, airfreight

On someone else's budget: Scrapped goods, extra units, extra transport costs for new unit, downtime, safety problems, low morale, missed delivery, lower long-term profit, unstable supply, customer dissatisfaction, idle trucks, management attention taken from the future

One effective way to get the entire organization's attention is to cost out an average breakdown by incident and by hour. Meet with your counterparts in accounting, sales, and operations, and start assigning estimates. You provide the maintenance numbers and have them provide the other numbers.

7.5.1 Case Study: The Real Cost of Breakdowns

Following is an analysis of the yearly breakdowns for Springfield Trucking at the request of Tony Willeby, the president of the organization.

YEARLY TOTAL COST OF BREAKDOWNS FOR SPRINGFIELD TRUCKING	
Number of Breakdowns	200
Number of Down Hours	1500

FIXED COSTS ASSOCIATED WITH BREAKDOWNS		
Item	Factor (Number of Incidents and Cost per Year)	Cost
Operator Lodging, Food, Expenses	80 x $100	$8,000
Extra Transport Cost	100 x $100	$10,000
Extra Costs from Core Damage	50 x $100	$5,000
Extra Costs from Damage to Associated Parts	25 x $250	$6,250
Towing Cost	50 x $200	$10,000
Extra Costs from Outside Vendor Parts and Labor	50 x $200	$10,000
Late Penalties in Typical Year	3 x $250	$750
Spoilage of Product	N/A	$0
Loss of Goodwill	N/A	$100,000
Cost of Overhead Associated with Original Assignment	N/A	$50,000
Total Fixed Costs Associated with Breakdowns		**$200,000**
Cost per Incident	$200,000 / 200	$1,000

HOURLY COSTS ASSOCIATED WITH BREAKDOWNS		
Item	Factor	Cost
Operator Idle Time – Total for Year	1500 hrs. x $25	$37,500
Extra Repair Time in Field Conditions – Estimated per Year	200 hrs. x $50	$10,000
Extra Travel Time for Mechanic	200 hrs. x $50	$10,000
Cost of Idle Units per Year	1500 hrs. x $25	$37,500
Extra Labor Needed due to Severity of Breakdown	50 hrs. x $100	$5,000
Total Hourly Costs Associated with Breakdowns		**$100,000**
Cost per Hour	$100,000 / 1500	$66.67

In this case, each breakdown that can be avoided saves the company $1000 per event and $66.66 per hour. Under normal conditions, a PM system can eliminate about 70% of breakdowns. You must balance the costs of a PM program against 70% of the costs of breakdowns. In Springfield Trucking's case, 70% of the annual $300,000 in breakdown costs comes to $210,000.

Accumulate your average number of breakdowns per year and compare 70% of that cost to the cost of the inspections, adjustment, and lubrication of a PM program. Assume that 70% of your breakdowns will be eliminated through an average quality PM system.

7.6 Remanufacture versus Replacement

Many types of vehicles can be economically remanufactured. Some major fleets run captive remanufacture shops. The US Postal Service has replaced the old Jeep mail vans with a unit manufactured by Grumman. One of the advantages of the Grumman van (in addition to larger capacity) is an aluminum body. The Postal service is figuring on a 15-20 year life for the aluminum body with 3-5 rebuilds (with new chassis).

Smaller fleets can also take advantage of this by ordering premium chassis in the first place to facilitate remanufacture. This is totally an economic decision. The cost of the remanufactured unit has to be lower than a new unit by a wide margin to make it pay.

Pepsi-Cola informs their bottlers that remanufacture costs are usually 1/3 of new costs and they can expect an additional five years of useful life. They look at capital availability and costs. During times of high capital costs, remanufacture becomes very attractive. On beverage trailers and route truck bodies, remanufacture costs 1/5 of replacement. Discussions about component remanufacture are included in the chapter on inventory.

Preventive Maintenance Systems and Procedures

To get the benefit of the installation of a PM system requires a commitment. In order to maximize the return of your investment in your equipment it must be kept in peak operating condition.

The PM approach is the long-term approach. Anything less than peak operating condition results in increased operating, maintenance, ownership or downtime costs. These costs vary slowly. Low overall costs of operation are the result of years of good maintenance policy.

Anyone can reduce maintenance costs for a few quarters by cutting back on PM inspections and the associated repairs. The temptation to do so is sometimes great because the consequences won't be felt for a couple of years.

Because of this temptation, and the length of time to get ROI, many fleet operations have either no PM system or only a minimal PM system. Some organizations inspect, lubricate, and adjust but don't feed back repairs to be scheduled unless they are clear and present dangers. Other organizations' PM task lists are on fixed intervals without any review of failure histories, service, or quality of drivers.

8.1 Benefits of PM Systems

Following is a list of benefits associated with the PM approach. If you are looking for arguments to support implementing a PM program, pick from the following list those that are most important to your business.

- **Reduced size and scale of repairs:** Detection of the critical wear point earlier in the cycle will reduce major repairs to minor repairs. A transmission might need a single bearing replaced early in the critical wear period and a complete re-build after a catastrophic breakdown. Maintenance (parts and labor) and downtime costs can be decreased as a result of smaller repairs.
- **Reduced downtime:** Units are kept in better condition. Reasons for downtime are tracked and the PM task list is adjusted to (if possible) solve these problems. Example: If you have downtime due to water in the fuel, you might have a PM task to check (and

correct) your in-ground tanks, or a task to inspect and drain the truck tanks.

- **Reduced core damage:** Early in the critical wear period the core is still intact. The core might be damaged by a total breakdown. PM will save parts costs by being able to reuse (or return) the core.

- **Increased useful life:** Better maintained equipment lasts longer. Consider the critical wear curve. Critical conditions that ruin a vehicle are managed before they have a chance to do their damage. Increased useful life reduces the ownership costs by dividing them over a higher utilization.

- **Increased driver/load safety:** Safety is a major factor of well-designed PM systems. All safety related systems (steering, braking, etc.) are checked regularly for critical wear. Reduced accidents decrease downtime, insurance costs, and increase morale throughout the organization.

- **Reduced road calls and associated lateness (time sensitive loads):** The PM inspector will catch 70% of critical conditions before they become breakdowns. Fewer breakdowns will directly reduce road calls and lateness. In addition to reductions to maintenance costs, there will be decreases to the downtime cost for the year. Additionally, responding to road calls is disruptive to jobs that are interrupted. Productivity increases in the shop with fewer interruptions for road calls.

- **Reduced overtime for responding to emergency breakdowns:** Fewer breakdowns translate into less overtime, reducing the labor component of the maintenance cost.

- **Increased vehicle availability:** PM systems keep units on the road by eliminating breakdowns and substituting short repairs and scheduled repairs for long repairs and unscheduled repairs. High availability increases the value of your service to your users and decreases short-term rentals. By eliminating the need to purchase additional units, ownership costs are lowered.

- **Reduction in spare units required:** Higher availability per unit means the fleet is effectively larger and translates to fewer spare units needed to serve the customer at the same level as before, resulting in savings from decreased ownership costs.

- **Increased control over parts:** Many of the PM parts can be assembled into kits, that can be pulled from stock in advance, facilitating efficient use of stock people. Quantities can be predicted from the PM schedule, allowing for more precise ordering and control.

- **Improved vehicle specification information:** The PM system tracks failures against all component systems, makes, and models of equipment. A review of these records will uncover good and bad specification choices. Improvement to specifications will reduce all cost areas during the life of the purchased unit.
- **Lowered overall maintenance costs:** Labor can be effectively scheduled and lost time can be reduced. Repairs are interrupted less often which increases productivity.
- **Lowered cost/mile:** All of the above result in lower cost per mile.
- **Improved identification of problem areas:** Correctly identifying problems helps to focus attention where it is needed.
- **Increased professionalism among maintenance workers.**

8.1.1 Additional Benefits of a PM Approach

- Your inspectors are your eyes and ears into the condition of your fleet. You can use their information on decisions to change your fleet make-up, change specification, or increase availability.
- Equipment has a breakdown curve. Once over the threshold, failures occur increasingly rapidly and unpredictably. Working lower on the curve adds predictability.
- Predictability shifts the maintenance workload from emergency firefighting, due to random failures, to a more orderly, even-handed, scheduled maintenance system.
- The frequency of user-detected failures will decrease as the inspectors catch more and more of the problems. Decreased user problems translate into increased satisfaction.

8.2 The PM Method

PM approaches to breakdown prevention are as old as the internal combustion engine. PM is widely accepted throughout the fleet world. The OEMs provide task lists based on miles, engine hours, even gallons of fuel consumed. These task lists are usually excellent starting points for a formal PM system.

In most fleets of any size, maintenance, repair and PM activity is monitored and organized under the CMMS (Computerized Maintenance Management System). The CMMS provides periodic reminders to bring a unit in for PM. The most sophisticated PM systems follow each unit using several parameters at the same time.

These systems will look at mileage, days, and, perhaps, quarts of add-oil. In this way, a unit will come up on days (say every six months), mileage (every 5000 miles) or when four quarts of add-oil is needed. In addition, these systems predict when the unit will reach milestones by extending recent usage into the future.

A PM System is divided into the PM activity, which is performed by the PM inspector, and the repairs to bring the unit back up to "like new" condition, which is the result of the inspection. These functions are separated because normally different people perform them. PM inspection generated repairs become the basis of scheduled work.

The Preventative Maintenance activity consists of the following on a task list:

- Predictive maintenance activities, including all inspections
- Scanning the vehicle's computer for fault codes
- Lubrication, oil changes
- Adjustments, cleaning
- Planned component replacements (PCR) such as belt/hose/filter changes
- Short or minor repairs up to 30 minutes in length
- Writing up any conditions that require attention (conditions which will lead, or potentially lead, to a failure).

A complete PM system consists of four additional essential factors:

1. Scheduling, **and actually doing**, repairs written up by PM inspectors.
2. Feedback of frequency of failures to PM inspection list.
3. Record keeping system to track PM, failures, and equipment utilization.
4. Continual training of inspectors, investigation, and investment in new inspection technology.

8.3 Costs of a PM System

- Investment (required prior to any benefit):
- Purchase of a manual or computerized record keeping system.
- Initial set-up of all of the files, lists of assets, parts kits.
- Engineering PM Task lists, frequencies, establishing standards.
- Initial inspector training.

- Modernizing and rehabilitating all equipment to PM standard (you are shooting for "like new" mechanical condition).
- On-Going Cost (paid every week, benefits will start to be felt):
- Labor and parts to complete inspection schedule.
- Staffing to record data in record keeping system.
- Staffing to analyze information collected by record keeping system.
- Labor and parts to complete scheduled repairs written up by inspectors.

8.4 The Language of a PM System

1. **BNF Equipment** – Equipment left off the PM system, left in the Bust 'N Fix mode.
2. **DIN Work** – Do It Now. Emergencies that interrupt on-going work and take less than one hour to complete.
3. **Emergency Work** – Work that was not scheduled a day in advance, which interrupts the schedule. Emergencies that take more than an hour to fix.
4. **Feedback** – Information from your individual failure history that changes task list.
5. **Frequency of Inspection** – How often do you do inspections? What criteria do you use to initiate an inspection?
6. **Inspectors** – The special crew that has primary responsibility for PM's.
7. **Predictive Maintenance Activity** – Any activity which looks both directly, and indirectly, at the condition of a unit/component and decides when, or if, a failure will occur. Predictive techniques include visual inspection, listening to the equipment, use of high technology such as engine analyzers, vibration monitors, etc.
8. **Scheduled Work** – Work that is written-up by an inspector. The scheduler will put the work into the schedule to be done. Sometimes the inspector finds work that must be done immediately which becomes an emergency or DIN (do it now) work.
9. **Task List** – List of activities that the inspector reviews.

8.5 Parts of a PM System

In setting up your PM system you need to answer the following questions:
- Which units do you include in the PM system?
- What tasks do you perform?

- How often do you perform PM activities?
- Who (skill sets) should do the PM tasks?

8.5.1 Units to Include

Compile a list of all of the assets (or units) that you are responsible for, including the following:

- Power Units including tractors, trucks, busses, cars, vans
- Trailers including vans, chassis, flat beds, bulk, tankers, specialized
- Reefer units on trailers, trucks, and containers
- Agricultural implements
- Rail Cars including bi-modal units, boxcars, flats, tankers, rail maintenance
- Construction equipment
- Stationary engines, generator sets, co-gen sets
- Overhead cranes, both indoor and outdoor
- Material handling equipment such as fork trucks, pallet trucks, piggy-packers
- Shop equipment of all types, shop air system, fluid systems
- Do you end up providing maintenance on building, machinery, etc.?

By the way, this is the same list you would compile to install your CMMS. If you have a CMMS, then you should have already done this step.

There are several criteria to determine if PM is worthwhile. All decisions on inclusion of items in the PM schedule are based on economic or safety justifications.

8.5.2 PM Task Lists

The task list consists of the items to be done: the inspections, the adjustments, the readings and measurements.

Sources of task lists:

- Manufacturers' OEM manuals
- OEM technical bulletins
- Government Regulations and State law (State DOT departments)
- Third party published shop manuals
- Trade Association recommendations

- Experience
- History

Other sources might include:
- Consultants
- New equipment dealers
- Third party service companies
- Your engineering department

The task list should be designed to perform basic maintenance, capture information about the future of the unit, or direct the attention of the inspector toward critical wear areas and locations. If you are inspecting an expensive component system, many inspections might go by without any reportable changes. Depending on the economics, you may want to continue to inspect to capture the change when it happens.

Failure experience feeds back into task list. You have to design standards to increase and decrease the number of tasks based on the failure history. Some organizations use the standard that if you don't get a reportable item every other PM then you are inspecting too frequently.

Add the cost of the task to the list using the following formula:

Cost of Inclusion = Cost of PM x Number of PMs per year

8.5.2.1 Types of Task Lists
- **Unit Based** – This is the standard type of task list. You complete the list on a single unit before going on to the next unit.
- **String Based** – This list is designed to inspect one, or a few, items on many units in a string. If the units are parked together in the yard or garage it might be easier to look at one item on each unit. The inspector's efficiency is higher when focused on one activity. Tire inspection or oil (fluid) checking are examples of good string based activities.

8.5.2.2 PM Levels

There can be many levels of PM. Most organizations use A, B, C, and possibly D levels. The lowest level (say the A level) would be an oil change, chassis lube, walk around, inspection on the lift and other more or less routine services.

The B level would be more in-depth and would always include all the tasks (or deeper ones looking at the same area) of the lower PM.

The C level might be a deep inspection with disassembly of systems and component replacement.

8.5.2.3 Specialized On-Condition PM

If you operate a computerized system, then you can afford to specify PM's that are relatively specific for a condition.

For example: A unit gets scheduled for PM based on high oil consumption (parameter = add-oil). Instead of doing a standard A level PM, perform a special PM based on oil items.

This task list could include:
- Oil analysis
- Check for oil leaks
- Compression test
- Examine filler cap
- Examine oil filter, oil filter seal
- Check oil pressure sender
- Check oil temperature sender
- Inspect anti-freeze
- Look at and analyze exhaust
- Other specific oil-related checks

In addition, since the unit is under your control and "on the lift", you might want to perform any other PM tasks that are almost due.

8.5.3   When to Perform PM tasks

One of the difficulties in planning frequencies of inspection is to know how often is often enough. This is the key to minimize total maintenance costs. If the frequency (expense) of the whole PM effort (all levels) is too high, the total cost goes up. If PM costs are too low, then breakdowns are high and total costs rise.

Each operation has an optimum level of PM activity based on the type of product they are moving, the criticality of the move, and the consequences of downtime. Other factors can include the service radius and the quality and competence of the driver. What you must do is continually adjust the frequency and task list to move to the center of the curve (the lowest overall cost).

Consider a snack distribution fleet. Usually the service radius is small and the criticality is low. The consequence of failure is low so the motivation for a rigorous PM system is also low.

On the other hand, consider an explosive hauler. The service radius is usually large, the criticality of the load is moderate but the consequence of a breakdown could be catastrophic.

Sources for inspection frequency include:
- OEM Owner's manuals
- D.O.T. regulations
- Third party shop manuals
- Insurance carriers
- Trade Association
- Experience
- State law
- History

The first source for inspection frequency is the OEM (manufacturer's) manual. Ignoring it might jeopardize your warranty. Usually the manual is concerned with protection of the manufacturer and limiting warranty losses. Following manufacturer's guidelines means you might be over-inspecting the equipment. The severity of the unit's service contributes to the frequency. The same unit needs different frequencies of PM if it operates on the Great Plains versus in the Rockies, or if it runs 24/7 instead of day time only.

Certain inspections are driven by law (State, D.O.T.). You have a certain amount of flexibility in the timing of these inspections (in Pennsylvania, you have a three month window to perform an Annual inspection). Consider scheduling them when a PM is also due. While you have the unit under your control you can also perform the in-depth PM to improve efficiency.

Your own history and experience are excellent guides because they include factors for the service that your equipment sees, the experience of your operators, and the level and quality of your maintenance effort.

The PM inspection routines are designed to detect critical wear and defer it into the future as much as possible. Since we cannot yet see the wear directly, the goal is to find a measure that is easy to use and is more directly proportional to wear. Traditionally two measures were used: utilization (miles, hours) and days.

For almost any measure to be effective, the PM parameter (such as miles, days, etc.) must be set for the unit (unique parameter table for each unit) or for the class (like units in like service).

8.5.3.1   PM Reset Concept

The great advantage of the computer in PM is that it can keep several clocks running for each unit based on information collected for other purposes. Most computerized PM systems track three or four of the measures (days, utilization, fuel, add-oil) for all units at the same time. A particular unit might get kicked for a PM on add-oil and another unit for fuel. Both units might still have miles left on the mileage clock.

The concept of reset states that doing a particular PM will reset the clocks on the lower PMs in that hierarchy. This has two applications.

In the previous example, after the PM is complete on the unit that was kicked for add-oil, all the counters or clocks for days, miles, etc. would be reset to start over. The slate would be clean on that PM.

The second application of reset is when you perform a high level PM, such as a B level PM, the counters on the A level (or lower levels) would also be reset. You don't want to rebuild an engine on Monday and have an A level PM on Thursday.

8.5.3.2   Clocks

There are other measures (in addition to KM, Miles, Engine Hours) that could be used to directly or indirectly capture wear or usage:

- **Gallons of Fuel** – This is an excellent PM measure. It reflects the overall utilization of the unit. You are probably already collecting gallons data for fuel consumption, fuel tax, and/or fuel security reasons. Fuel is a more accurate measure for critical wear than mileage or even engine hours. Fuel consumption includes the variability of rough service, excessive idle time, and engine/component wear (increased friction). In a heavy duty unit, you might initiate a "B" level PM every 2000 gallons (instead of 10,000 miles).
- **Add-Oil** – This is a direct measure of the condition inside the engine. Engine wear and condition are directly proportional to oil consumption. You might schedule the same "B" level PM after 12 quarts of add-oil.

A complete set of parameters for a PM (shown below) might be to initiate an "A" level PM if **any** of these parameters is exceeded:
- 90 days or 5000 miles
- 250 hours or 1000 gallons of diesel
- 6 quarts of add-oil

8.5.4 The PM Inspector

PM is double-sided. On one side is basic maintenance including oil changes, checking belts, chassis lubrication, etc. This is relatively lower skilled work.

On the other side is inspection. August Kallmeyer (a retired AT&T maintenance expert) says this about inspection, "A successful PM program is staffed with sufficient numbers of people whose analytical abilities far exceed those of the typical maintenance mechanic".

You need analytical individuals who will be able to detect potentially damaging conditions before they actually damage the unit. Look for individuals with the following traits or skills:

- The PM inspector should be trained in techniques of analysis and in the use of the most current inspection tools.
- The PM inspector should know how to review the unit history and the class history to see specific problems for that unit and for that class.
- A PM inspector is pro-active (whereas a mechanic is re-active). In other words, the inspector must be able to act on a prediction rather than react to a situation.
- Because of the nature of critical wear, the more competent the inspector, the earlier the deficiency will be detected. The early detection of the problem will allow more time to plan, order materials, and help prevent core damage.
- PM inspectors should be full time inspectors, segregated from the rest of the maintenance crew. If the inspector's primary job is repair and renovation, then the PM system will suffer.
- In some large operations, the PM group receives a pay premium and may have a different style of uniform.

In any facility the PM inspectors should represent 10-15% of the whole crew.

One of the legacies we fight is the old concept of the grease monkey mechanic. Through the PM effort, and other approaches, we need to increase our professionalism. In other repair fields, such as computer repair and copy machine repair, professionalism is a job requirement.

8.5.4.1 PM Repairs

Should the same individual perform the PM and complete the repairs? From a productivity viewpoint, it is always more productive if the worker is multi-skilled and can do the whole job without moving the vehicle. In addition, the communication gap is eliminated by having the PM person also complete the corrective work.

The other issue to consider is the completeness of the inspection. Will an inspector write up everything knowing they will be the ones who will have to do the work? According to fleet expert, Ron Turley, this has not been shown to be a problem. In fact, in most cases, the mechanic is more conscientious about writing up corrective actions. If the mechanic does miss something, usually it is something that didn't really need to be repaired in the first place. Turley goes on to state that the best situation occurs when the PM inspector/repair mechanic is also responsible for the vehicle on a long term basis.

8.6 Implementing a PM System

	STEPS IN IMPLEMENTING A PM SYSTEM
1.	Create a PM task force. This group includes shop people (including the shop steward in union shops), supervisors, operations supervisors, data processing representatives and (if possible) engineers.
2.	Decide on the goals of the task force. Set objectives.
3.	Get training in computers for members of the task force if they are not computer literate. Include typing training. Secure access to computers and word processors, spread sheets, e-mail, and any relevant organizational level networks or systems. If there is a maintenance system (CMMS), start extensive and ongoing training.
4.	Get generalized maintenance management training for the entire task force. This will save time and effort by laying groundwork so that they share a common language and create a new vision of maintenance. A Maintenance Management Certificate program would be ideal.
5.	Identify the maintenance stakeholders (anyone impacted by how maintenance is conducted). Analyze their needs and concerns. Look at each group and see how they contribute to the success of the organization. Include production, administration, accounting, office workers, tenants, housekeeping, legal, risk management, warehousing, distribution, and clients. Analyze how the proposed changes benefit each group.
6.	Write first drafts of the measures or benchmarks that will be used to evaluate the PM system's performance. These measures will be revised as the process continues.

7.	Begin to draft the SOP (standard operating procedures) for the PM system. This document will be revised many times over the first year. Search out and incorporate any related SOPs.
8.	Inventory all equipment to be considered for PM. This includes everything that you are responsible for.
9.	Select a system to use that will store information about equipment. Select forms for PM generated ROs and check-off sheets.
10.	Have task force members, contractors, or shop personnel, complete data entry or preparation of equipment record cards. Rotate the job so that everyone in the department has experience with the system before you go on-line.
11.	Complete a daily audit of all data typed into the system. Have a highly-skilled individual review this data. Don't skimp on this step or the system will be full of garbage.
12.	Consider using vendors to replace the hours lost on the floor by individuals engaged in data entry. It is essential to build a critical mass of expertise in the system.
13.	Select people to be PM inspectors. Incorporate their input into the next steps. Use inspectors to help set-up the system.
14.	Determine which units will be under PM and which units will be left to break down. Remember that there is a real cost associated with including any item in the PM program. If you spend time on PMs for inappropriate equipment, you may not have time for the essential equipment.
15.	Use the following formula to calculate the cost of including each **piece of equipment**: Cost of Inclusion = cost per PM x number of PMs per year
16.	To decide which units to include in the PM system, apply the following rules to each item: • Would failure endanger the health or safety of employees, the public, or the environment? • Is the inspection required by law, insurance companies, or you own risk managers? • Is the equipment critical? • Would failure stop production, distribution of products, or complete use of the facility? • Is the equipment a link between two critical processes? • Is it a necessary sensor, measuring device, or safety protection component? • Is the equipment one of a kind? • Is the capital investment high? • Is spare equipment available? • In case of failure, can the load be easily shifted to other units or work groups? • Does the normal life expectancy of the equipment without PM exceed the operating needs? • Is the cost of PM greater than the costs of breakdown and downtime? • Is the cost to get to (view or measure) the critical parts prohibitively expensive?
17.	Is the equipment in such bad shape that PM won't help? Would it pay to retire or rebuild the equipment instead of PM it? Schedule modernization on units requiring it. Plan to retire BNF units, if possible.
18.	Select which PM clocks you will use (days, utilization, energy, add-oil). A clock is designed to indicate wear on an asset. Clocks on items in regular use, or subject to weather, are usually expressed in days. An irregularly used asset might be better tracked by usage hours. Some equipment, such as construction equipment, is best tracked by gallons of diesel fuel consumed, because the hour meters are frequently broken.
19.	Set up task lists for different PM levels and different classes. For example, lists for trailers every month, then every year, with three different lists "A, B, C" for tractors. Factor in your specific operating conditions (mountain, desert, marine), skill levels, operators' experience, etc.
20.	Categorize the PM tasks by source (recommended by Ron Moore of RM Group). Categories might include regulatory, calibration, manufacturer's warranty, experience, insurance company, quality, etc. This will be of great benefit when you look back to see which ones to eliminate or change.

21.	Decide what Predictive Maintenance technology you will initially incorporate. Train inspectors in techniques. Even better, provide the information and a budget to the task force and let them pick the technology. Most equipment should be rented before buying. Inexpensive training is available from most vendors and distributors.
22.	Assign work standards to the task lists for scheduling purposes. Observe some jobs to get an idea of timing. Let some mechanics time themselves and challenge them to re-engineer the asset to reduce PM time.
23.	Provide the PM inspector with the following items to perform the tasks: • Actual task list (usually a work order) with space for readings, reports, observations. • Drawings, performance specifications, pictures, where appropriate. • Access to unit history files, trouble reports. • Equipment manual. • Standard tools and materials for short repairs. • Any specialized tools or gauges to perform inspection. • Standardized PM parts kit. • Forms to write up longer jobs. • Log type sheets to log short repairs.
24.	Determine frequencies for the task lists. Select parameters for the different task lists.
24.	Engineer all the tasks. Challenge yourself to simplify, speed-up, eliminate or combine tasks. Improve tooling and ergonomics of each task. Always look toward enhancing the worker's ability to do short repairs after the PM is complete.
25.	Implement the system, load the schedule, and balance hours. Extend the schedule for 52 weeks. Balance to actual crew availability. Schedule December and August lightly, or not at all. Allow catch-up times.

8.6.1 PM Case Study – A Mining Fleet

This case study involves a mining fleet in Alaska that includes twenty haul trucks, dozers, shovels, and loaders. Alaskan fleets generally turn on their vehicles in October/November and keep them running until spring.

A consultant was called in because the fleet was experiencing what they considered to be excessive breakdowns while doing what they claimed to be first-class, complete and up to date PM.

Even before seeing the fleet in person, the consultant began thinking about what could account for breakdowns while doing PM.

The list included:
- PM inspectors that are incompetent and/or inexperienced.
- Inspectors that are looking at the wrong things (by not following the tasks).
- Inspectors that are looking at the wrong things (by following the tasks).
- Task lists that are bad.
- Frequency that is wrong (too infrequent).
- Equipment that is too small for the work (PM can't put iron into a machine).

- Equipment that is junk and in breakdown life cycle.
- Not actually doing the PMs (pencil whipping the forms).

Upon arrival, the consultant toured the mine and maintenance facility.

In speaking with the PM inspectors and touring their shop, it was clear they knew their business. This was a group of old-timer pros. The consultant couldn't imagine them pencil whipping the PMs. They were using the appropriate Cat task lists with their recommended frequencies. The equipment was being used in ways the consultant had seen before at other similar mines. The equipment seemed somewhat junky but not inappropriately so. The condition of the equipment was a small question mark. Superficially, the consultant eliminated most of the potential problem areas in the first few hours.

The next step was to interview the different groups, along with the operations people. Maintenance complained that operations never gave them the equipment long enough; while operations complained that the trucks were taken out of service too long. The disposition between the two groups towards each other was not great. Although not unusual, it is a symptom of something more going on.

Sometimes the numbers tell the story. The CMMS (which seemed to be well used) showed work accomplished ratios of:

- 15% for PM work
- 9% for Corrective work
- 40% for Breakdown work
- 36% for Project work

The ratios were calculated from direct labor hours for the last six months. Other periods showed similar ratios. These ratios were a bit unusual. But what they indicated would take some digging to reveal.

There are few rules in managing fleets. As long as the number of fatalities is low, and the production targets are met, there will be little external oversight. How work progresses through a shop is unique to each company.

The consultant diagrammed the work flow through the shop, the flow through the CMMS, and how work was allocated. The work flow showed the problem. All the other findings supported the conclusion.

In short, the PM inspectors were conscientiously doing the PMs at the recommended frequencies and coming up with a list of necessary corrective items. The corrective items involved deterioration found by inspection before failure. To do the corrective list completely would require the truck to be down for weeks. Operations were releasing the trucks for PM since PM took only a day or two. But they were not releasing the trucks for the corrective work because it took too long. Occasionally, a truck would be so bad that the maintenance department would commandeer it and fix everything. It would be off the line for weeks.

Items found by PM inspectors were already in a death spiral and were going to break eventually. The whole idea of PM is to detect this state of deterioration and fix the component or system before breakdown. In this fleet, the items called out by the inspectors failed before the shop could get to them!

The equipment was in deteriorated condition. Without the corrective work, all the systems of the trucks were in various states of deterioration.

What would account for the weird work accomplished ratios? Normally, you would expect ratios of 15% PM, 55% Corrective, 15% Breakdown and 15% Projects.

In this fleet, the high level of project work was made up of maintenance workers tied up for months doing major refits and rebuilds of large equipment (which is usually considered capital improvement rather than maintenance). Frequently, this kind of work is done by contractors. This meant that even if operations relented, and gave up the trucks for full Corrective actions, there were not enough mechanics to do the work.

This was a tough situation. After years of operating this way, repairing this amount of deferred maintenance would cost millions of dollars. A business decision would have to be made whether to pursue the corrective work (perhaps gradually) or set up to be the best firefighting maintenance group north of the Arctic Circle. If commodity prices stayed low, and the proven reserves of the mineral are limited, there might be no choice but to become fire fighters.

8.7 Higher Level Predictive Maintenance Techniques

Predictive maintenance is a natural part of the PM system. Predictive maintenance is undertaken from the PM task list. Any activity designed to observe and guess when a failure will occur is predictive. All inspection (visual, sound) is basically predictive. This section will concentrate on predictive techniques that require some level of higher technology.

In the beginning, we said that the ideal situation would be to be able to peer inside your components and replace them right before they fail. Technology has been improving significantly in this area. Tools are becoming available that can look into an engine, thread through an exhaust system, or detect a bearing failure weeks before it happens.

Predictive Maintenance is a maintenance activity geared to indicating where a piece of equipment is on the critical wear curve. Much of the inspection activity on the PM task list is predictive in nature. We make the distinction in this section to discuss only techniques that require special tools, instruments, or materials. All of the techniques discussed herein should be integrated into and controlled by the PM system.

In some cases, fleet managers have borrowed tools from other fields such as facility/plant maintenance, oil pipeline inspection services, racing, and medicine. Techniques include spectrographic oil analysis, infrared temperature scanning, magna-flux, ultrasonic imaging and CCD inspection. In all cases, most metropolitan locations have service companies to perform these services. Other instruments such as exhaust pyrometers, strain gauges, temperature sensitive tapes and chalk can also be useful.

8.7.1   Oil Analysis

By far the most popular technique to predict current internal condition and impending failures is spectrographic oil analysis. This is based on the fact that different materials give off different characteristic spectra when burned. Oil is primarily a lubricating media and secondarily a heat exchange material. It acts somewhat like the blood in your body. Almost anything happening in the engine will eventually show up in the oil. The results are expressed in PPT, or PPM (parts per thousand, parts per million).

The lab, or engine manufacturer, usually has baseline data for types of equipment that it frequently analyzes. The concept is to track trace materials over time and determine where they come from. At a particular level, experience will dictate an intervention is required (a re-build, re-manufacture, etc.). Oil analysis costs $10 to $25 per analysis. For larger fleets, it is frequently included at no charge (or low charge) from your motor oil supplier.

A report is generated with a reading of all the materials in the oil and the "normal" readings for those materials. In some cases, the lab might call (e-mail or fax) the results in so that you can finish a unit, or capture a unit before more damage occurs. For example, if analysis shows antifreeze in the oil, then a breach has occurred between the cooling and lubricating systems inside the engine (or contamination outside the engine). This engine is in trouble and shouldn't run.

As another example, an increase from 4 PPT to 6 PPT for Chromium probably indicates increasing liner wear. This could be tracked and checked on in the regular inspections.

Oil analysis includes an analysis of the suspended, or dissolved, non-oil materials including:

- Babbitt
- Copper
- Lead
- Aluminum
- Molybdenum
- Silicon
- Titanium
- Chromium
- Iron
- Tin
- Cadmium
- Nickel
- Silver

In addition to these materials, the analysis will show contamination from:

- Antifreeze
- Bacteria
- Water
- Leather
- Acids
- Fuel
- Plastic

Additionally, the oil analysis gives you a view of the condition of the oil itself:

- Has the oil broken down?
- What is the viscosity?
- Are the additives still available?

Consider oil analysis a part of your normal PM cycle. Since oil analysis is relatively inexpensive also consider it:

- Following any overload or unusual stress.
- If sabotage is suspected.
- Just after (or better yet, before) purchasing a used unit.
- Following a bulk delivery of lubricant to determine its quality and if bacteria is present.
- Following a rebuild, to baseline the new engine and for quality assurance.
- After service with severe weather such as flood, hurricane, or sandstorm.

Oil analysis is just as effective on gear oils, transmission and hydraulic systems.

There are some variations on the theme of oil analysis. Oil analysis deals with particles up to 50 microns. On larger particles, ferrography is used to determine the shape and size of the wear particles. Large and irregularly shaped particles indicate abnormal wear.

On even larger particles, chip detection is used. In its most primitive form, magnetic plugs are placed on the oil sump that the PM mechanic then inspects for unusual amounts of metal. More sophisticated detectors will set off an alarm if they detect metal particles over a certain size.

Oil Analysis firms exist in most major cities, and can be found via the Internet, especially cities that serve as transportation centers. These firms will prepare a printout of all of the attributes of your motor, hydraulic, or transmission lubricants.

8.7.2 Temperature Measurement

Since the beginning of the internal combustion engine, temperature has been an important issue. Most equipment contains temperature gauges for oil and/or water. Temperature is the single greatest enemy of oil and components of the engine and transmission. Advanced technologies in detection, imaging, and chemistry, allow us to use temperature as a diagnostic tool.

Today, technology exists that can photograph heat rather than reflected light. Hotter parts show up as redder (or darker). Changes in heat will graphically display problem areas where wear is taking place, where a breach has occurred in an exhaust system, or where there is excessive resistance in an electrical circuit.

Direct applications include testing of external engine parts such as water pumps, heads, exhaust systems, and radiator/external cooling parts. Readings are taken as part of the PM Routine and tracked over time. Impending failure shows up as a change in temperature.

Temperature detection can be achieved by infrared scanning (video technology), still film, pyrometer, thermocouples, other transducers, and heat sensitive tapes and chalks.

On larger stationary engines and turbines, temperature transducers are included for all major bearings. Some packages include shutdown circuits if temperatures get above certain limits.

8.8 Increased Vehicle Availability and PM

Earlier, we stated that ownership costs were 25% of the fleet dollar spent and through proper management you can reduce ownership costs by 10–20%. One major opportunity for improvement is in the area of vehicle availability and utilization.

In a 250 unit fleet, if you can increase availability 3% (from 90%–93%) that is equivalent to expanding the fleet by seven vehicles. With low availability, those seven vehicles are already funded for depreciation, insurance, permits, and a license, even if they spend most of the year parked against the fence.

With increased availability, the units can be used for expansion, trade-ins, or analyzed and used to replace higher cost units. Together, several of the highest cost units in your fleet could be used to fund a new unit. This constant winnowing of your fleet will help keep your costs in line.

Availability and Utilization Targets:
- 98% cars, light trucks
- 94% Medium duty trucks
- 90% Heavy-duty equipment

Your PM plan fits into this calculation as the means by which to get from your present utilization levels to the target levels and higher.

The availability of units when your user department/group needs them is part of your mission. In many organizations, units are kept as spares because needed units are down. In the fire-fighting oriented atmosphere of a repair shop (not a maintenance facility) there is little idea of availability.

PM decreases emergency breakdowns. Lower breakdowns mean that vehicles are more often available when the user group needs them. Higher availability also means that fewer spare units are required to maintain the same effective fleet size. Fewer vehicles mean lowered ownership costs.

8.8.1 Case Study in Vehicle Availability

In a city in the North Eastern US, a fleet of garbage trucks is being reviewed by a consultant. The review was initiated by the fleet administrator. The administrator is trying to figure out how well the fleet is doing after having hired a maintenance manager from the private sector.

The new maintenance manager has accomplished what seems like a miracle. After six months, he has increased the availability of the garbage truck fleet from 49% to 60%. The consultant suggested an informal benchmarking study to compare fleet availability among other entities. The maintenance manager asked that the benchmarking be restricted to other cities (not private companies like Waste Management or BFI which operate in the same region).

Following are the results:
- New York – 85%
- Philadelphia – 81%
- Our city – 60%

The truth is that an increase from 49% to 60% is a significant achievement. However, there is still a ways to go.

As availability increased, other problems started showing up. The fleet had 72 routes to run with 100 units. As they closed in on 72% availability, they found that operations would not pick up the trucks as they were completed. The trucks were left in the yard because operations didn't have enough crews to run all the trucks that were being made available.

8.9 Planned Component Replacement (PCR)

We use the standard that inspectors and the PM system should generate 70% of the ROs. This means that 30% are generated by operators, emergencies, and DIN type jobs.

In some industries, notably the airlines and nuclear energy industry, 30% for non-scheduled repairs is too high. The implications of emergency failures in those industries are too grave for 30% non-scheduled work.

PCR was developed as an option on the PM task list. The novelty of this option is the elimination of failure because components are removed and replaced after a pre-determined number of hours but before failure. The components are then returned for inspection, rebuilding, and remanufacturing. The result of this strategy is controlled maintenance costs. A PCR program results in higher maintenance costs with correspondingly higher reliability.

Fleets are beginning to realize that costs of unscheduled downtime are sufficiently high to justify these more advanced views of the maintenance activity. Since the component is replaced before failure on a scheduled basis, PCR offers the fleet the following advantages:
- Core damage is less likely.
- Replacement is scheduled to avoid downtime (replace the component when the unit is not needed) and reduce overtime (from breakdowns).

- Tools can be made available on a scheduled basis to reduce conflicts.
- Manufacturers' revisions, enhancements, and improvements can be incorporated more easily.
- Rebuilds in controlled environments by specialists are always better than the same rebuilds "on the floor" by general mechanics.
- The rebuild (since it is scheduled) can be used for training of newer technicians.
- Spare components can be made available on a scheduled basis that can minimize inventory (rather than waiting for breakdowns which are known to clump together).
- Since the component is replaced, breakdowns become infrequent, availability goes up, and the atmosphere becomes more regular.
- In a successful PCR plan, management will take the time to look at any failures that do occur and seek ways to avoid failures of this type in the future. Some options are better quality lubricants, better skill in repairs, design review, and specification change.

8.9.1   Case Study – PCR

A pick-up and delivery fleet (this would be a class of vehicles – like vehicles in like service) has been investigating reoccurring failures with an eye toward increasing fleet reliability. To investigate, a report called Component Life Analysis was used. This report showed the elapsed utilization between failures of the same systems. For example, all the braking system problems were coded under VMRS (see chapter on Systems to Aid Fleet Managers) system code 13. The system calculated the elapsed mileage between brake problems (as well as all the other systems).

The trucks carried small packages, with a full truck carrying thousands of small packages. While the direct lost revenue if the packages were late was substantial ($15,000 and up), the loss of customer good will was worth ten or more times the lost revenue.

Excessive failures of the starter motor, across the fleet, in this class of vehicles was discovered. The MTBF (Mean Time Between Failures) was about 93,000 miles. They also found that the standard deviation (see a complete explanation in the chapter on Statistics) was small, meaning that most of the starters were failing pretty close to the 93,000 miles.

The best thing to do would be to determine if there was a solvable root cause of the failures (called root cause analysis). Their investigation revealed no causes that they could see across the class.

The solution was to bring the trucks in for a Planned Component Replacement of the starter motor. It turned out to be cost prohibitive to bring the trucks in for just the starter motor replacement. However, it also turned out that the trucks were already being brought in for a 75,000 mile PM check up. It would be easy, and add minimal extra cost, if the starter PCR were added to that PM. The 75,000 mile change of the starter would insure that very few of the starters would fail catastrophically and potentially impact the pick-up and delivery operation.

8.10 Case Study – Comparing Breakdown, PM with PCR and PM with Inspection Programs

Tom Duvane has hired your consulting company to evaluate several different maintenance programs. They currently use P&M Truck Leasing for full service leasing on their power units. They are interested in having the Springfield Central Garage provide maintenance services for their reefers. Tom feels that the outside work will improve his productivity and provide profit for the company.

COMPARING BREAKDOWN, PM WITH PCR, AND PM INSPECTION PROGRAM COSTS ON 50 REEFER UNITS	
Vehicles	50 TK Reefer units mounted on Great Dane Reefer Vans
Utilization	2500 hours / year / reefer
Belt Failure Rate	Mean = 575 hours SD = 175 hours
Failure Rate	2500 (utilization hours) / 575 (belt failure rate) = 4.35 per unit 50 (units) x 4.35 (failure rate) = 218 failures for the fleet
Cost	**Per Event**
$285.00	Nonscheduled (emergency) failure
$85.00	Scheduled replacement
$25.00	Inspection cost plus scheduled replacement cost when belt is replaced
$2.50	Administrative cost per inspection
$20.00	Administrative cost per repair incident
Breakdown Program	
fleet failure rate x (cost per nonscheduled failure + administrative repair cost) = Total Cost of Breakdown Program	
218 x (285 + $20)	**$66,490**	Total Cost of Breakdown Program
PM with PCR	
mean failure rate hours - 1 SD = PCR Interval	
575 - 175	400 hours	PCR Interval
Use 400 hour PCR interval to identify 84.9% of failures	

(utilization hours / PCR interval) x number of units = CR Incidents		
(2500/400) x 50	313	PCR Incidents
fleet failure rate x (100% - 84.9%) = Emergency Incidents		
218 x 15.1%	33	Emergency Incidents
PCR incidents x (scheduled replacement cost + administrative repair cost) = PCR Incidents		
313 x ($85 + $20)	$32,865	PCR Cost
emergency incidents x (cost per nonscheduled failure + administrative repair cost) = Emergency Cost		
33 x ($285 + $20)	$10,065	Emergency Cost
PCR cost + emergency cost = Total Cost of PCR Program		
$32,865 + $10,065	**$42,930**	Total Cost of PCR Program
PM with Inspection		
Based on PM interval of 200 hours to catch 96.5% of failures		
(utilization hours / PM interval) x number of units = PM Incidents		
(2500/200) x 50	625	PM Incidents – Number of Inspections
fleet failure rate x 96.5% = PM Replacements		
218 x 96.5%	210	PM Replacements
fleet failure rate x (100% - 96.5%) = Emergency Incidents		
218 x 3.5%	8	Emergency Incidents
PM incidents x (administrative repair cost + administrative inspection cost) = Inspection Cost		
625 x ($20 + $2.50)	$14,062.50	Inspection Cost
PM replacements x (scheduled replacement cost + administrative repair cost) = Scheduled Replacement Cost		
210 x ($85 + $20)	$22,050	Scheduled Replacement Cost
emergency incidents x (cost per nonscheduled failure + administrative repair cost) = Nonscheduled Failure Cost		
8 x ($285 + $20)	$2440	Nonscheduled Failure Cost
inspection cost + scheduled replacement cost + nonscheduled failure cost = Total Cost of PM Program		
$14,062.50 + $22,050 + $2,440	**$38,552.50**	Total Cost of PM Program

In this case, the PM with Inspection program costs the least, but this method requires the most maintenance hours which could be a problem in some companies.

Measuring Worker Productivity

9.1 Work Sampling

Imagine taking random snapshots of your maintenance mechanics. You would find that, at the instant of the snapshot, a percentage of your crew is involved in non-productive activities. Work sampling is a formal technique to evaluate the activities of your maintenance work force. Most supervisors do their own informal version of work sampling as they walk around. This section introduces a formal methodology.

Using work sampling management can secure facts about the operation without watching everyone all the time. It is a systematized spot-checking method where different observers under the same conditions will get the same results. Under certain circumstances, work sampling can be more accurate and reliable than continuous observation, as done in time study.

In the November/December 1986 issue of IPE (and verified by the author several times more recently), investigators from Emerson Consultants, Inc. published the results of work-sampling 35 typical industrial maintenance departments including all types of maintenance.

AREAS AND AMOUNTS OF LOST TIME EACH DAY	
Bargaining Agreement Time Losses – rest breaks, meals, wash-up, including normal plant practices such as get ready, etc.	78 minutes
Idle Time – no job assignment, unsanctioned, avoidable delays	44 minutes
Excess Personal Time – eating, talking, smoking, drinking, and resting, in excess of the provisions of the bargaining agreement	35 minutes
Late Starts and Early Quits at beginning and ending of shift, before and after lunch and breaks	21 minutes
Travel to and from job assignments, transporting materials, tools, or the unit itself for service	77 minutes
Waiting for materials, tools, or for the unit to be serviced	22 minutes
Getting job assignment, instructions	21 minutes
Picking up and putting away tools	25 minutes

The study concluded that there were 323 minutes per day of non-productive or marginally productive time. During the normal 8-hour (480 minute) day, the average maintenance worker spent 157 minutes on the job, using job related tools. 95% of the improvements to labor productivity come from these 323 minutes of lost time per day.

Work sampling is important because:
- You must know how deep your organization's individual pool of lost time is.
- You need to baseline your shop as a starting point.
- After you install some labor productivity improvement you need to be able to measure its effect (is the pool of lost time getting shallower).

We acknowledge the pioneering work of L.H.C. Tippett, who introduced the field of work sampling in England in 1934. Robert L. Morrow, of NYU, introduced work sampling here in a paper to the ASME in 1941. More recently, A. Kallmeyer has done extensive teaching on the subject. Don Nyman and Emerson Consultants applied the ideas to managing maintenance. This section is adapted from the work of these men.

9.1.1   Case Study in Informal Work Sampling

A consultant was asked to teach a fleet maintenance management class for the supervisory staff and management of a municipal government fleet near Los Angeles. Like most municipalities, they had a wide range of equipment from large road equipment to small lawn mowers. The classroom was located in a mezzanine with wall-to-wall windows overlooking the main shop. You could see into seven repair bays from the window. There were curtains over the windows to reduce distractions.

When the consultant reached the section on worker productivity, the shop superintendent took issue with the low level of productivity stating that his people worked six hours a day or better (75%). Everyone in the class immediately had an opinion about how conscientious their people were and that the data in the study did not relate to them.

The consultant offered to conduct an experiment right then. Using the random time table (see later in this section), random times were selected and put in order. When one of the times came up, the class would stop, open the curtains, and take note of what those workers, visible through the windows, were doing. The shop leadership had to promise not to act on the information until the class was over and they understood the importance of the results.

The class waited excitedly for the first random time. At the appointed time, the curtain was opened…and no one was seen to be visibly working!

Of course, that does not mean that no one was working. People could have been working in the yard, road testing, etc. All that could be said was that no one was working in the shop in the bays.

Over the course of three days, at most, two of the seven bays were working at any one time. The superintendent was ready to run out and yell at everyone. But before any yelling occurs, there should be an understanding of why people are not working in the bays. It is important to see what this sampling data means.

9.1.2 How to Use the Results of Work Sampling

Work sampling itself is not a problem solver. It is more of a problem finder. If properly planned, it will give very definite indications of what should be done. For example, your study may show that excessive time is spent waiting for materials at the issue window in the morning. Some judicious planning might allow the parts room person to pull standard jobs the night before, when the window is quiet. Then, in the morning, 75% of the mechanics could begin work immediately.

Once problems have been isolated, effective means can be used to improve the situation.

There are several classes of reasons that categorize why employees are not at their posts, and work sampling will help uncover which classes dominate. Observations at many shops over the years indicate that 90% of the lost time is due to gaps in management, and management incompetence, and only 10% from employee laziness or employee problems.

Usually, when confronted with this information, supervision moves immediately towards punitive approaches. Yet most of the problems are actually imposed on the maintenance workers by the management system and management decisions.

Take a common example:

Many shops save money by limiting the number of common tools available. In a shop with this management strategy, you might only find one rolling oil drain. What happens when a second person needs to change oil? That management decision forces the employee to wait, slowing down both people.

A few other examples:
- Unit is not available when the employee goes to work on it
- Employee is waiting for someone with special skills or knowledge
- Missing tools (time spent looking)

- Next job's repair order is not in the hands of the employee when they finish their current job
- Employee spends excessive time entering data into a slow, unfriendly CMMS
- Broken tools (time spent fixing or improvising)
- Inefficient shop layout with long walk to parts area or tool storage crib
- Stockroom doesn't have part (and unaware of it)
- Anything that stands in the way of a productive job

In the final analysis, the productivity of the workers is a function of having what they need when they need it. In order to complete a job, the mechanic must have access to the following:

- A work order or request to do a particular job.
- A person with the right skill(s) that is physically and mentally able to perform the job.
- A helper if necessary.
- Custody and control of the asset.
- Correct parts, materials, supplies, and consumables for the job.
- Any special tools that are required.
- Adequate equipment for lifting, bending, drilling, welding, etc.
- Safe Job steps that are either already known to the mechanic or in a manual or planning sheet.
- Identification of hazards and mitigation, including Personal Protective Equipment (PPE).
- Safe access to assets (decontaminated, cooled down, etc.), safe work platforms, and humane working conditions.
- Up to date drawings and wiring diagrams and other information.
- Proper waste handling and management.
- Knowledge of what tests to perform before returning the asset.

If anything is missing, the mechanic will have to either search around for it, work around it, or improvise. All of which reduce the ability of the worker to do the work they were hired to do.

9.2 Conducting a Work Sampling Study

Before you start:
- Define the problem you want to solve. You may want to start with a general study to see how much time is spent in the general

categories. At a later date, you might want to study waiting time related to the Parts Room.

- Plan a study that will address the problem at hand. Assign people to the study.
- Review the study with your people. They may not like the idea but should be informed of their responsibility to contribute to an efficient operation. They should be shown how high efficiency will improve their job satisfaction.

Once the study is underway:

- Randomness is the first key to the whole study.
- Vary your routes through the maintenance facility to increase randomness and surprise people. You can also train other people to conduct the study and randomly vary the observer. Consider training supervisors to conduct the study in other departments.
- Select random times from the random timetables. Plan 4-6 tours per day. Allow a reasonable time between tours. If the selected time runs into lunch or another established break, stop the tour. If the random time occurs during lunch, then skip that tour.
- The second key to success is impartiality. Do not prejudge what you see or mix in outside factors (for example, thinking that someone is a good or bad worker). You are trying to analyze the system, not the people. Do not make assumptions about what the people were doing or what they are about to do, only note what they are currently doing. It is important to record a person's activity before they see you.
- Use one tally sheet per shift. Use a separate sheet for each observer.
- Fill out the random times and random tour routes in advance. Enter them in the observation time and observation route rows. Conduct the tours using the routes indicated at the times indicated.
- Record the number of people at work that day. This is the number of observations you will make each tour. Enter that number in the available manpower row.
- During each tour, record the observation elements (what is being done) in the column of the specific observation time. You can use tally marks as you walk through. One element per observation per person per tour.
- If you observe crew members working in the wrong area or on the wrong job, mark the observation based on where they are at that moment. Don't judge where, or what they are doing, just if they are working.

- At the end of each shift, total the observations and transfer to the recap sheet.

\multicolumn{5}{c	}{**WORK SAMPLING TALLY SHEET**}
Organization:	Date:	Observer:		
Location of Study:	Number of Employees at Work:			
Tally Marks	Tour 1	Tour 2	Tour 3	Tour 4
\multicolumn{5}{c	}{**Productive Activities**}
PM				
Scheduled Work				
Emergency/DIN Work				
\multicolumn{5}{c	}{**Non-Productive and Marginally Productive Activities**}
Bargaining Time				
Idle Time				
Travel Time				
Waiting Time				
Getting Instruction				
Clean-up/Set-up				
Total Observations				
Unobserved Employees				

*Total Observations + Unobserved Employees = Employees at Work

9.2.1 Use of the Random Timetable

The random timetable is used to select times to tour your facility and make observations. The table can easily be used in a number of ways. The most common would be to start in the upper left and move across the table. You can also use the times vertically, or on the diagonal. Start filling in times from the table on to your observation sheet. Reorder the times from random to time order when you enter them on the observation sheet. At the end of each day, take note of where on the random table you end, and start there on the next day. Any time that is too close to a previously selected one can be discarded and used on a later day's tour.

9.2.2 Random Time Table

9:40 10:40 1:45 7:00 11:00 9:30 7:20 8:50 9:55 2:45 10:40 1:45 7:00 11:00 9:30 7:20 8:50 9:55 2:45 5:00 3:15 10:35 5:15 8:35 3:20 11:50 1:55 12:35 6:40 9:10 9:05 9:30 8:00 12:45 5:15 10:00 7:30 7:30 4:05 9:10 4:35 7:45 5:20 11:50 2:05 6:15 6:45 3:55 11:25 8:30 5:15 8:50 12:40 5:10 8:40 11:20 6:05 11:55 3:55 10:05 10:30 10:15 7:50 3:40 8:45 2:30 3:20 4:30 12:35 9:30 12:25 4:50 7:25 6:50 12:10 2:40 10:30 9:20 5:05 8:05 7:15 6:10 4:30 12:15 10:50 7:40 6:30 11:25 9:20 2:05 8:35 6:40 3:00 8:45 3:20 9:45 11:30 7:25 2:30 9:05 2:50 9:35 12:10 2:05 6:35 12:50 2:30 10:35 11:25 7:05 1:55 8:00 11:40 6:00 2:35 7:25 6:10 11:45 9:10 11:35 4:10 2:35 7:35 4:40 1:45 9:00 8:35 10:05 7:30 4:15 12:40 1:10 9:35 10:10 3:55 6:20 1:55 3:05 9:55 7:15 11:00 12:50 11:20 10:00 5:25 9:00 1:30 5:30 9:30 8:10 12:00 4:05 5:40 10:55 1:05 7:40 6:10 8:10 3:00 12:30 11:10 7:05 6:35 1:25 4:45 9:30 4:00 2:20 6:55 5:00 2:45 2:05 4:50 8:55 7:35 9:30 2:45 9:05 2:10 9:30 8:20 11:05 8:00 5:15 10:45 5:50 1:00 9:35 1:15 11:55 12:30 2:50 9:30 5:20 11:05 8:20 2:45 7:05 2:00 5:00 11:00 11:45 2:50 3:10 2:25 7:40 3:35 11:10 1:50 12:55 1:40 5:10 7:20 10:10 6:55 9:05 3:00 7:40 3:10 7:00 6:10 12:45 8:55 3:50 8:05 9:25 6:05 5:10 6:20 11:00 3:50 12:40 3:20 4:20 5:45 9:55 10:30 7:35 10:50 3:50 6:50 6:00 1:05 11:20 1:35 11:15 5:40 7:00 4:20 7:50 11:15 7:35 10:10 7:50 9:55 2:05 12:25 1:10 3:05 9:45 4:20 10:05 6:00 11:45 2:15 4:30 6:45 8:25 5:25

Work Standards

There are three major reasons to develop, use, and track work standards. The first reason is to be able to evaluate individual mechanics. Comparisons of individual work to work standards will uncover and, perhaps more importantly, prove who needs training, re-assignment, or a new job.

The second reason to use standards is to develop a scheduling system. Scheduling repairs and PM activity requires estimates of time to be effective.

The last major reason to use work standards is to be able to accurately predict when a unit will be returned to your users. Accurate estimates improve your customer service and increase the value of the service you perform for your users.

Effectively scheduling a Maintenance facility requires some idea about the amount of time a job is going to take. The era of $5 to $6 per hour shop labor is gone. Labor is the major component of the maintenance dollar. Establishment of labor standards is the first step in the control of labor.

10.1 Types of Work Standards

10.1.1 Flat Rates (FR)

Flat rates are the old name for labor standards. They are published by the OEMs and third party publishers. The flat rates are used to reimburse the OEM for warranty claims. They are rigorously looked at and developed through time study of skilled mechanics using a full set of tools. Activities such as diagnosis may be separated from the actual repair (see what is included). Road testing is usually not included. Air tools are usually assumed.

Skilled mechanics should be able to meet or exceed flat rate work standards. The flat rate for a repair may be short if the vehicle is old, has had very rough service, the components are rusted in place, or the mechanic is not fully trained. These rates often use specific tools that might or might not be available in your shop. If rusted bolts or studs break during the repair, then even a highly skilled mechanic is likely to miss the flat rate with good reason. The only way to know if a flat rate book is good for your equipment and your shop is to observe several repairs (see RE below).

FR are often used in incentive schemes. Some automobile dealers charge the customer by the flat rate for the job and then pay the mechanic a percentage of the flat rate. Repeat repairs are handled on the mechanic's own time (or other arrangement).

FR are a good place to start as long as they are compared to the other standard techniques. If you start using flat rates to estimate repair times then 100% would be the average productivity rate for skilled people. Schedules based on flat rates should be de-rated by a percentage based on the relationship between the flat rate and the experience of your crew (use FR with RE below).

10.1.2 Historical Standards (HS)

Just about all computer systems track the time it takes to do every job. Some systems can evaluate all of the repairs of a particular type on a particular class of equipment. These standards can be useful since (unlike the flat rate) they factor in important variables such as the actual condition of your equipment, the skill level of your workforce, shop practices, the layout of your shop, and your tools and equipment.

The disadvantage of historical standards is the accuracy of the data collection and the definition of the tasks completed. For example, the RO might read Fix Steering. The mechanic might have to check tires, alignment, and frame damage before accepting that there is even a problem with the steering. All of these activities are lumped together in the historical standard.

Historical Standards (HS) should be based on:
- Class of Equipment – like equipment in like service is grouped together
- System/Assembly/Part – what exact part was replaced/fixed (for example – left front brake shoe)
- Work Accomplished – was this an adjustment, replacement, inspection – what work was actually done
- Labor Hours – should reflect 1 or 2 person crews
- Location of the repair – conditions may vary from shop to shop and repairs on the road take significantly longer

- Factors such as mechanic, repair reason, and labor or parts costs are not considered in historical labor standards.

Schedules based on HS include a full amount of lost and wasted time. When an employee is talking, eating, or not at his/her job, the time is still charged to the repair. The time will find its way into the historical standard. HS are a less adequate place to start a performance evaluation scheme. Flat rates (vs. HS) don't usually include lost time but they also don't take into account the conditions and skills of your crew.

10.1.3 Reasonable Expectancy (RE)

The third method (RE) can be used as a check, as a factor times the first two (FR, HS) methods, or by itself.

Reasonable Expectancy (RE) is a work standard based on observation. The definition of RE is a reasonable, observable amount of work in a given amount of time. Because of the inherent inclusion of lost time (in the HS) and not taking into account your fleet and environmental conditions and skill levels (in the FR), we feel that the RE is the best way to accurately schedule a repair facility or evaluate a mechanic (particularly a repetitive repair).

The concept of reasonable is important. By observing several actual people doing the repair you have a good idea of how long it should reasonably take. No speed-up is needed to improve productivity. You will improve productivity by recapturing time lost in non-productive and marginally productive activities.

This is not a system of time study. It is far more informal and easier to develop and maintain. The comparison is micro versus macro analysis. With RE's, you are looking for the condition of having an employee at work, you aren't necessarily concerned with the individual's technique or skill. Time study looks at the individual technique to evaluate its effectiveness. Hundreds of times more information is required to properly time study a repair compared to observation and assignment of an RE for the same repair.

10.1.3.1 Instructions for RE Observations

Preparation for establishing RE's:
- Determine what activities you want to observe. Start with repetitive repairs and PMs. Review your repair history and list the most frequent repairs.
- One of the keys to success is capturing the whole job. For example,

if moving the oil drain from the last place it was used is always part of the oil change, be sure to include this time.
- Discuss the study with all affected employees. Sell the concepts of improved customer satisfaction (your own user group), improved quality of life for shop people, improved ability to recognize excellent performance, and include a discussion about bad actors, those who make more work for everyone.

When you start:
- Fill in the top half of the observation sheet before you start the observation. Recap all observations by the end of the day while they are still fresh in your mind.
- Position yourself so you can see the entire bay where the repair is going to take place. Try to limit observations to 20-30 minutes. Break-up large jobs if convenient or assign another observer.
- Use discretion in recording observations. No stop watches. Do not carry on a conversation with the employee. Answer questions briefly.
- Be on the scene before the activity starts and after it is completed. Try to observe different people doing the job. Three or four observations for each repair are usually enough if there is not a great unexplainable variance in the times.
- When recoverable lost time is observed, document in detail what occurred. If, for example, you observe talking, note with whom and the content of the discussion (is it job related). Don't act immediately on lost time. This study is not to be used as a whip, but, rather, to determine a reasonable expectation of a day's work. Later, after your schedule is installed, is the time to act on lost time issues.
- Make an entry on the observation sheet with a start and finish time every time a function changes. Function changes include travel, setting-up, fetching materials/tools, actually working, and both recoverable and non-recoverable lost time. Record and document all data relevant to the observation. Decide and record if the circumstances of the observation were normal.
- Substantial changes in tools, shop layout, or procedure require new observations.

10.2 Using Standards to Track Employee Activity

The measurements discussed in the Measurement of Fleet Expenditure include tracking your mechanics to standards. One performance measure compares the RO hours against the standard hours for the same repairs. The second measure compares standard hours to payroll hours for an overall measurement of labor effectiveness. Any of the three standard hour systems can be used.

10.2.1 When There is No Standard

Leading fleet maintenance consultant, Ron Turley, says many jobs can be simply measured by counting the number of bolts and multiplying by three minutes each. This includes removing the bolt, replacing the bolt, doing the mechanical part and set-up for the job. At one end, such as replacing a starter with three bolts, the estimate is too short. At the other end, where there might be 100 bolts (and air tools are used) the estimate might be too long. In the middle, 10-50 bolts, it works pretty well.

10.2.2 When the Standard Cannot be Met

If an employee does not meet the standard, it is a minor issue. Intervention should only occur after long observation. Most of the reasons that employees cannot make the standard are related to management decisions. However, the following reasons do require intervention:

1. The employee is not sure about the job steps.
2. The employee has received bad information.
3. The employee is sure about the steps but works in an unusual or inefficient way.
4. The employee does not spend enough time actually working (busy locating tools, parts, manuals).
5. The employee does not spend enough time actually working (busy visiting with friends).
6. The employee does not spend enough time actually working (sick – in the bathroom excessively).
7. The employee is tired, sick, or infirm (temporary).
8. The employee is not strong enough, tall enough, or smart enough to complete the work.
9. The employee is avoiding doing a day's work.

You can see that the first three items are clearly lack of skill, training, or knowledge, and could be training issues. Item 4 is most closely related to the management system. Even item 5 can be related to the management system (though not always). Items 6 and 7, if they are temporary, can be overlooked. If they persist, the employee might be temporarily reassigned to lighter duty. Item 8 (assuming that training will not change this situation) is a problem if you cannot reassign the person and don't have alternative duty available for them. Item 9 is famous but actually happens less often than most people surmise. In most cases, attitude problems are secondary to some other problem.

The root cause in all these cases can usually be traced back to four causes. Observation can sometimes uncover or distinguish the root cause. The four root causes are:

1. Lack of training, skills or knowledge.
2. Bad management system supporting mechanic, bad shop layout, bad lighting, bad supervision, inadequate tools.
3. Lack of aptitude as in not strong enough, or tall enough to reach certain things or even not smart enough for the job.
4. Bad attitude.

10.2.3 Use of the Three Standards

Tom Devine wanted a method to measure the productivity of his mechanics. He obtained the Flat Rates (FR) from the Chilton Labor Guide.

Ann Moore, a co-op student from the local engineering school, calculated the Historical Standards (HS) from repair orders for the same repairs over the last year. She made Reasonable Expectancy (RE) observations in the month before the study.

Tom chose Joe Dillon, a responsible but slow employee, to evaluate. Joe didn't like paperwork and constantly complained about it. Tom figured if the standards worked with Joe, they should work for anyone else.

The week of October 1 through 7, Joe Dillon had 32 regular time repair order hours and was paid for 40 regular time hours. He completed eight chargeable jobs. The remaining eight hours were spent on non-chargeable activities (indirect activities) such as cleaning up and fixing the grounds department lawn mower (should have been on a RO).

EVALUATING PRODUCTIVITY WORKSHEET		
Work Standard Hours (WSH)	The sum of the standard hours for each job completed.	
Performance Effectiveness	WSH/ROH	WSH/RH
	Performance Effectiveness	
	RO Hours (ROH)	Regular Hours (RH)
	32	40
Work Standard Hours (WSH)	**Calculations**	
Flat Rate (FR) – 28	28/32 = 87.5%	28/40 = 70%
Historical (HS) – 31	31/32 = 96.8%	31/40 = 77.5%
Reasonable Expectancy (RE) – 25	25/32 = 78%	25/40 = 62.5%

Tom noticed several things from the study. The flat rates were somewhat faster than the historical performance at his main garage. He also looked at the RE's and couldn't see why anyone couldn't make them.

As he observed Joe Dillon over the next few days, he saw that Joe worked at a pace which should easily make the RE's. When Tom looked closer, he saw that Joe spent an excessive amount of time filling out Repair Orders. Tom discovered that Joe could barely read and write English. Remedial classes solved the problem and made for a happier employee.

Tom concluded that the REs presented the greatest opportunity for improvements. They seemed to point out where problems were and also seemed intuitively fair. These advantages outweighed the higher costs associated with them. He also realized that in the future, RES and FRS used together might be the easiest to use on a day-to-day basis. He could see adjusting the FRS by the REs to minimize the number of observations.

Staffing

There are several approaches to the coverage of the week. As mentioned in a previous section, it is advantageous to repair the vehicles when they are not in demand. In fact, if you run vehicles during the day (such as a local delivery fleet) and repair them at night, you can get by with fewer spares (you save about 3% of the total vehicle census). Weekend work should not be necessary if the entire second shift is available. If, and only if, it is needed, weekend work can be covered by overtime or staggered starts.

Even if demand is daytime only, and the bulk of maintenance is carried out at night, a small day shift is important. They are there for occasional emergencies and to keep larger jobs moving along in a reasonable time. The day shift will also insure special parts and outside work is expedited.

The issue is the premium pay for evenings and weekend work. If you don't move the week around, you will be paying premium pay for the extra time. There are several strategies that have been utilized by repair facilities to address this issue. Following is a chart with some rules of thumb and calculations to show the practical impact of various shift schedules. It is based on the experiences and research of Mike Brown of New Standard Institute.

SHIFT SCHEDULES CHART

Hour Shift	Available Ratio	Cost Ratio	Fatigue Ratio Week 2	Fatigue Ratio Week 3
8 Hour 5 Day	1.00	1.00	1.00	1.00
8 Hour 7 Day	1.00	1.21	0.90	0.80
10 Hour 5 Day	0.95	1.16	1.00	1.00
10 Hour 7 Day	0.95	1.51	0.85	0.73
12 Hour 5 Day	0.83	1.40	1.00	1.00
12 Hour 7 Day	0.83	1.60	0.80	0.64

The **Available Ratio** relates the work time after subtracting established meals and breaks. So an 8-hour shift is 1.0 and a 12-hour shift adds 4 hours but also a meal break.

The **Cost Ratio** only adds in factors for time and a half and double time where appropriate (USA standard—this might be different in other areas).

The **Fatigue Ratios** are rules of thumb for people working continuously. In any shift pattern, a two-day weekend rests the employee enough so that he or she can start fresh.

11.1 Staggered Starts

One strategy is to stagger the work week. People start on different days and work 5–8 hour days or 4–10 hour days. In the second case, employees have three days off (usually a good incentive). In some jurisdictions, they are also paid overtime for the two hours over the eight hour shift. In one shop, the crews worked three 12-hour shifts one week (for 36 hours) and four 12-hour shifts the next week (for 48 hours). This gave them the advantage of being able to cover 24 hours with only four shifts. Of course, you can see from the Shift Schedules Chart how costly this would be.

Another strategy is to stagger start times. In some shops where manning a second shift is a problem, it is often easier to have a few people start early and a few start later. These people can work on all the small jobs to maximize the number of vehicles on the street.

11.2 Staffing Roles

There are a variety of roles to fill in a fleet maintenance facility. The number and qualifications for each of the roles depend in part on the size, age, location and makeup of the fleet.

Mechanics – Skill level depends on the depth of the work undertaken. There should be a range of ages and skill sets so everyone can develop and make a contribution.

Vehicle Washing, Fuelling and Jockeying – This role includes driving vehicles to, and picking up vehicles at outside shops, running for parts, and minor vehicle service (such as topping off oil). Generally, a washer can wash a dirty truck in about 20 minutes and an automobile in about 10 minutes (both depend on the level of mechanization and the quality requirements). It takes 6–7 minutes to fuel a vehicle. If the driver is standing around, they should do the fuelling. For long haul trucks, a fuelling set-up with two nozzles attached to one meter (called a master-slave) will increase efficiency.

Operators and Drivers – The drivers or equipment operators might be highly skilled individuals. In any case, they have to complete a pre- and post-trip inspection. These are mostly safety items but some maintenance items can and should be added. Supervisors should observe operators to insure the complete inspection is done. There could be an annual test to certify operators for the inspection.

Parts Window, Restocking, Receiving – Whether or not there is a parts window person, there is a requirement for someone to receive, stock shelves, cycle count, send out for repair, tidy up the parts room, place reorders and find stuff that has gotten lost. Generally there is one of these employees for every twenty mechanics.

Purchasing Parts, Fuel, Tires, Purchasing Outside Services – A fleet uses 40% or more of its budget on parts. When tires and fuel are added in, purchases become the largest dollar consumer. Frequently, a representative of Purchasing will be assigned to the fleet and do all the buying and negotiating with both service and part vendors.

Supervisors and Lead People – Generally there should be about one supervisor for every 10-12 workers. For small shops, with fewer than seven mechanics, the supervisor also works interruptible jobs. The supervisor has a multitude of roles including job assignment, scheduling, disciplinarian, lawyer, psychologist, lion tamer (just kidding), advocate, and a few others.

Clerks – The trend is toward fewer clerks and more direct data entry. With bar codes, portable computers, PDAs, and other devices to speed the process, this is very doable. The newest generation of worker is computer literate so the training burden is minimal. A large shop should have one or two clerks for all the clerical tasks that seem to fall through the cracks. In a small shop, the clerk might do double duty as a buyer, contract officer, or parts stocker.

MECHANICS NEEDED TO MAINTAIN VEHICLES BY CLASS	
Type of Unit	Service Units Per Mechanic +/- 10%
Passenger Cars and Pick-up Trucks – Moderate Service	75
Passenger Cars – Heavy Service (police, taxi, etc.)	35
Light Delivery Vans (Fed Ex)	40
Medium Trucks (UPS)	30
School Busses	25
Transit Busses	12
Heavy Diesel Trucks – Pick-up and Delivery	20

MECHANICS NEEDED TO MAINTAIN VEHICLES BY CLASS *(cont.)*	
Type of Unit	Service Units Per Mechanic +/- 10%
Heavy Duty Trucks – Long Haul	18
Garbage Trucks – Urban	10
Concrete Trucks – Construction Sites	12
Oil Field Service Vehicles (mobile derricks)	5 (or fewer)
Mixed Municipal Fleet (everything soup to nuts)	30
*These ratios are guides	

11.3 Active Supervision

Active supervision can improve productivity by 15%. The question is what is active supervision? Active supervision occurs when the supervisor spends substantial time on the shop floor helping workers solve problems. As strange as it might sound, on the psychological level, the supervisor might have to be both mother (nurturing and supportive) and father (strict and tough) to members of the crew.

The supervisor is seen by some as the key player in removing the obstacles that get in the way of the mechanic having a productive day. In the section on work sampling, we discussed how the level of productivity of workers can be a function of having everything they need to get their work done. The supervisor is the tool who can provide this. They are one step ahead of the jobs on the floor making sure everything is ready for the next job on the line.

Active supervision is broken into several dimensions:

Ongoing Performance Monitoring – The supervisor knows how long each job should take and checks it periodically throughout the day. A 4-hour job should be half done by break. When the jobs fall behind, the experienced supervisor provides the best intervention. In some cases it might be logistical help, tool help, or information about how to proceed. In some cases, it might be a kick in the pants! In other cases, the supervisor will hang back if wrestling with the job is important for training.

Paperwork Compliance Goon – The accuracy of all analysis is derived from the work order. If the work order is complete and accurate then decision making is dramatically easier. The supervisor is always auditing paperwork and returning it when it is deficient. He or she should always look at work orders on the floor and insure entries are being made contemporaneously (at the same time) as the activity.

Teacher and Mentor – The supervisor should be either training him/herself or directing the training of members of the crew. Everyone has areas that they are better at and areas that they are worse at. The easiest crew to schedule is one where everyone can do everything. The effective supervisor should be reviewing the schedule every day and look for training opportunities. These can be formal training sessions or letting the trainee help an experienced hand.

Quality Control Officer – The supervisor is responsible for the overall quality of all work performed in his or her shop. Where there are quality issues the supervisor determines the cause of the problem whether it be lack of knowledge or skill, lack of aptitude, lack of adequate physical strength, dexterity, bad attitude, lack of the right tool, lack of the right part, some inadequacy in working conditions, inadequate time, preoccupation brought about by a problem outside of work or other reason. The supervisor works with the worker to solve the quality problem. If the problem is with the company or system (bad conditions, lack of tools or parts) he/she should tackle that also. If the worker has a problem with external issues, the supervisor should mentor them or find them help in the organization.

Safety Officer – The supervisor should intervene any time an employee or visitor performs an unsafe act or is in the shop without personal protective equipment. The supervisor is the champion for safety and makes sure the shop is safe.

Tidiness Cheerleader – The shop must be kept clean for safety, efficiency and morale reasons. All clean-up for individual jobs should be part of and charged to the individual job. The supervisor should arrange for periodic clean projects to keep the whole area and the yard tidy.

Psychic – The supervisor must see the problems in the future and fix them in the present.

Shop Design

Before you undertake the design of a new maintenance facility (or an extensive redesign of an existing facility) have an answer to the following questions:

- How many pieces of equipment will be serviced by this facility?
- What types and sizes of equipment (Heavy duty, diesel, gas, lawn mower, etc.) are there?
- What are the demand hours for the vehicles?
- Is this a major dispatch center also (are the vehicles housed, loaded, fueled here)?
- What categories of work are you expecting to do in this facility (light repair, engine rebuild, truck body mounting, welding, body work, etc.)?
- What and how many work shifts are expected?
- How much reliance is there on vendors and outside shops?
- Do you have vehicles that stay out on the road for excessive periods?
- How many mechanics do you need?
- How many bays do you estimate that you'll need?
- What is the projected future of this division, area, and facility?

12.1 Turley's Rules of Shop Design

12.1.1 The Shop

The best design is to have the trucks face head in toward a central aisle. Located in the central aisle are the parts area, supervisor complex, rest room and tool room. This gives you the shortest walk to all the essentials.

Use penetrating sealant on the concrete before you start using the shop. This will greatly simplify daily sweep up.

12.1.2 The Bay

If the bay is against a wall, allow 8' in front of the vehicle measured from open hood or tilt forward cab. Allow 8' between equipment for normal maintenance. This gives you a 16' wide bay. If you work on cars and small trucks, limit the width to 14'. Some shops use 20' wide bays with the idea of utilizing the extra space for storage. In actuality, the area seems to fill with junk and becomes a hazard and impediment to productivity. When two bays are next to each other they share an aisle (especially when you follow the rules of tool placement). If there is a wall, you must allow an extra 6'. Add a few feet for body work bays. Isolate body work bays from the shop to exclude dust and fumes.

Overhead doors should be remote operated (if possible) and 14' by 14' in truck bays, 10' by 10' in automobile bays. Consider doors with windows to take advantage of daylight.

Each bay should have an air hose connection at the head and tail. Overhead hoses for engine oil and antifreeze should be located at the head. Lighting should be in the aisles, between the bays, and in the central aisle. Paint the walls and ceiling with reflective paint to increase light and work level. Pick a lamp to give as natural light as possible to facilitate color recognition. Consider skylights for added natural light.

Mechanic tools are located on either side of the central aisle facing toward the bay. Work tables, special storage, and cabinets are also located along this aisle. Everything that cannot roll should have a place off the floor. The mechanic assigned to the bay should be responsible for keeping it picked up. They might also be responsible for keeping it clean (or a janitor will deep clean the floors after they are picked up by the mechanic).

12.2 Managing Tools and Equipment

Look closely at your shop operation and see the tools that are used daily or even hourly. Tools like oil drains, and tire pressure gauges have a very fast payback.

Tools like air impact guns can cut time and fatigue tremendously.

Electronic and computer based tools are expensive and increasingly essential.

Shadow boards are great when you keep them up.

The question is at what point does it pay to invest in having an extra tool available or in having a tool available for every mechanic?

Tools are expensive. The natural tendency is to spend as little as possible to get the job done. The result is that tools are scarce. This seems consistent with the idea that the best way to save money is to not spend it in the first place. There is a fallacy in this thinking. The mechanic looking for a tool, waiting to use a tool, or standing in line to take a tool out of the crib has a cost. There is a real (generally not calculated) labor cost to this waste. It turns out that this cost can be quite high.

Ron Turley did some calculations based on some assumptions. He used a three year payback period for tools and a 2.5 burden factor (times direct wages) for the burdened labor rate. Based on a three year payback, the dollar savings from issuing an individual tool to each mechanic was equal to the wage rate per hour times the minutes per day saved times 30.

The math is as follows:

Burdened Labor Rate x Average Number of Work Days / Minutes per Hour = Cost per Minute

$$\frac{(hourly\ rate \times 2.5) \times 660}{60} = Cost\ per\ Minute\ for\ 3\ years$$

where...
- Burdened Labor Rate = Hourly Rate x 2.5
- Average Number of Work Days in 3 years = 660
- Minutes per Hour = 60

You can then calculate:

$$hourly\ rate \times \frac{(2.5 \times 660)}{60} = 27.5 \times hourly\ rate$$

and round 27.5 to 30.

In today's climate, you would push for a two year (440 days) or shorter payback and use 15-20 times your hourly rate. Reworking the formula:

$$hourly\ rate \times \frac{(15 \times 440)}{60} = 110 \times hourly\ rate$$

Supplying each bay with an oil drain, saving 10 to 15 minutes per day, with an hourly rate of $20, would save you $22,000 per year:

$$\$20 \times 110 = \$2,200 \times 10 = \$22,000$$

12.3 Waste Oil

One area where there has been an excellent amount of work done is in waste oil burning systems for garages in cool and cold areas. Technology has improved to the point that government agencies have approved the products to reduce overall pollution.

These furnaces burn used motor oil, transmission oil, #2 oil (heating fuel and diesel), and hydraulic oil. In the US, they can be used to burn oil generated from your facility and also from do-it-yourselfers in your area.

In the US, the EPA approves this strategy because the waste stream is disposed of a where it is produced. It mitigates the danger of transport of hazardous materials.

The heaters must be able to filter the oil and vent to ambient air. Per regulations, in the US, they can't exceed 500,000 BTU per furnace. A gallon of waste oil has about 140,000 BTU. In typical installations, the payback is 18 months or less.

Planning and Scheduling

13

Use the following assessment form to get an idea of where you are in regards to planning and scheduling, where improvements are needed, and where to go to make those improvements.

PLANNING AND SCHEDULING ASSESSMENT FORM	
Effective Storeroom and Stock system	Investment, Culture
Effective PM program so that work is identified before failure	Culture
An up to date Maintenance Technical Library with an extensive library of vehicle manuals, flat rate books, interchange guides, etc.	Investment
Timely reporting of potential problems by drivers (to provide lead time for planning)	Culture
Good use of a CMMS or manual system	Training, Culture, Investment
Meaningful written comments on completed jobs by supervisors and technicians	Culture
Complete equipment Repair History	Investment, Culture
Thorough PM inspections	Training, Culture
Thorough failure analysis when breakdowns do occur	Training, Culture
Good relationships with vendors	Culture
Open dialogue with drivers and dispatchers on troublesome equipment	Culture
Existence of overhaul and rebuild capabilities	Investment
Good workmanship by craft personnel and an attitude toward solving all quality problems	Training
Good use of repair technology (ongoing training relationship with both OEMs and local tech schools)	Training

13.1   Shop Scheduling Tips

The first items scheduled are the units still in bays. Use substantial effort to solve whatever problem is keeping them from being finished. You can put a unit back on the street unfinished if the defect will not interfere with safe operation (such as a broken heater in the summer).

Rule: If possible vehicles that are started are worked on until they are completed.

Never start a job you can't finish (due to parts, tools or outside services). Identify all parts and other resources needed before starting.

- When units come in the door for any reason (PM, breakdown, corrective), review the unit history and see if any there is any other work due. If a unit comes in for a breakdown and is due for PM the following week, consider scheduling the PM while have control of the vehicle.
- Most of the schedule will come from PMs that are scheduled. They will constitute 10-15% of your work load and create an additional 45-55% (corrective items).
- Reserve parts by pulling them out of stock and putting them in a staging area (some places use plastic totes). One shop had a wall of old bus lockers that they used. Parts were pulled, put into totes, and slid into a locker. The key was put into a plastic envelope with the work order. Start the job when everything is there.
- Is there a day-of-week effect? If so, then some of your demand is known by the day of the week.
- Look outside. The weather will immediately influence the schedule for that day.
- Overtime should be the result of a short term inequality between the demand for services and the resources available. It should be known about well ahead of time. If there is an emergency requiring overtime then mechanics can work on routine work to fill in the time, finish the shift, or while waiting for the unit.
- Control your vacations with annual sign-up sheets. In one facility, people signed up for vacation at the beginning of the year and then again a quarter at a time. The order was rotated so that everyone had a chance at first choice. The number and skill sets of the people on vacation at any one time were regulated.
- Limit yourself when assigning more than one person to any job unless absolutely necessary. Two or more people slow the job down. Of course, safety sometimes dictates when two people must be used. Never have only one person in the shop.
- Supervisors should show up randomly if they are responsible for off shift work.
- Keep overlay between shifts to a minimum. The supervisors should be overlapping and finding out where each job is and passing that on to the crew member.
- Run as few shifts as possible. Three shifts are tough to crew and manage (and are usually less productive).

13.2 The Work Side

- All known repairs identified and repair orders written. Sort the repair orders by priority and add estimated times.
- All PMs scheduled with estimated times.
- Be sure that all knowable parts are available and any specialized tools are lined up.
- Get all vehicles that you plan to work on lined up (this might be done on a daily basis for the next day).

13.3 Short Interval Response Schedule (SIR)

The SIR schedule may be one of the most powerful tools for recapturing lost shop time. Short Interval Response is a control concept. Its concepts include:

- To control the whole, control the parts. You must manage at the time and at the level where you have a useful impact.
- The key is monitoring the operation repair-by-repair, mechanic-by-mechanic, hour-by-hour and operation-by-operation.
- If repairs are falling behind schedule (called a schedule miss condition), supervision is notified within a short interval. This way the situation can be corrected while they can have an impact on that job.
- If the schedule miss condition continues then higher levels in the organization are notified.
- Work is quantified. A reasonable amount of work is expected each day. Workers are freed from a hurry atmosphere one day and a kill time atmosphere the next. Work quantities are based on FR (Flat Rate standards), RE (Reasonable Expectancy standards) or some combination of both. Management's expectation of how much work is a reasonable day's work is given to the mechanic in advance. Intermediate goals are identified and checked by the supervisor.
- An inevitable by-product of this approach is the uncovering of many hidden operations problems.

13.3.1 Why SIR Scheduling Techniques Work

There are several reasons why the SIR schedule works and why it improves the quality of life in the organizations that use it in an enlightened way. These reasons are rooted in psychology, and one reason at least goes back to the early pioneering Hawthorn studies at the Chicago Hawthorn plant of Western Electric*.

In encapsulated form, the study found that workers at Western Electric responded with increased productivity to attention from management. The classic study concerned the effect of lighting on productivity. When lighting was increased, productivity went up. When lighting decreased, much to their surprise, productivity also went up. After many experiments, it was found that attention from management was the critical factor, not the lighting.

For a complete account see Fritz Roethlisberger and W.J. Dickson, Management and The Worker (Harvard Univ. Press 1939)

The SIR Schedule organizes management's attention and applies it where there is a problem. The attention naturally causes the productivity to go up. Additionally, higher levels of management are able to apply their problem solving skills to a real problem at hand.

The SIR schedule also works based on the psychology of perception. People have a strong desire for completion (see the work of Wolfgang Kohler and the original gestalt psychologists). The SIR schedule uses that desire by issuing each job with a clear near-term goal (either the job completion or a milestone). Completion in this sense is completion of the job. People will work a little harder if they feel the possibility of completion of a goal is do-able. Additionally, most people want to feel successful in front of peers and supervisors. Completion of scheduled jobs allows this.

Frequently, the reason a schedule miss is occurring is related to the mechanic not having a critical piece of information, such as special tools, techniques, general inexperience, or a possible material substitution. The supervisor, having been informed, can intervene and get the job back on schedule.

The SIR schedule can help a medium quality supervisor become a good supervisor. A good supervisor is already doing many of these things informally.

SIR schedules are not appropriate everywhere. In some cases, a subset of the schedule is indicated. Operations where there are less than ten maintenance workers would have difficulty justifying the full system due to the cost of the required planner/scheduler (usually the service writer). These smaller fleets can implement a subset of the SIRS strategy. Organizations that run their shop from a visible schedule and use informed estimates receive great benefits.

In organizations that are anti-staff (will not let you add a staff person even if you can prove greater savings in maintenance workers), the SIR technique is difficult to install. This approach requires significant investment before any savings are made. Organizations must be willing to make the investment. Top management must be interested and committed or the project is doomed to fail. SIR schedules require this commitment from the highest levels in the organization because it will bring about a fundamental change in the way your organization does business.

The system highlights areas where mechanics cannot do their job due to a problem outside their control and/or problems stemming from the old way of doing business. These previously hidden (or unpublicized) problems suddenly come to the foreground:

- Mechanics pulled off to work on non-productive activities
- Stock Room contributing to schedule miss conditions regularly
- Failure to put equipment back into service when promised
- No ability to get control of vehicles when needed for schedule
- Quality becomes a big issue
- Lack of cross training causing clashes of resources
- Failure to meet RE (reasonable expectancy) results in the immediate conclusion that it must be wrong
- Inability to handle emergencies, excess absenteeism, and unusual peak load conditions

13.3.2 History of SIR

The SIR schedule is both a technique and a way of thinking. It was first developed in the 1930's. Legend has it that Montgomery Ward needed a system to coordinate pulling and packing so that all of the items on a mail order would hit the packing table at one time. Each order might have anything from shoes to barbecues to lawn mowers to baby clothes. The engineering consultants were supervised by Alexander Proudfoot. His company, formed in the 1940's, is still in the business of improving productivity.

They called their development the Short Interval Schedule (SIS). For the last several decades, they have installed the system in a variety of organizations. They specialize in industrial, clerical, housekeeping, or commercial environments. Traditionally, SIS schedules have had some difficulty in dealing with non-repetitive tasks like maintenance.

Maintenance was considered a problem area for SIS because the scale of savings for the effort was not as high as production or other repetitive areas. Frequently, PM activities were included in the SIS because machine downtime could be a significant expense. Today, it is recognized that maintenance may be one of the major areas left for significant savings. It is has also been recognized, through work sampling analysis, that maintenance has relatively large amounts of unproductive and marginally productive time.

An excellent text on the subject states the dilemma of SIS and maintenance clearly, "The biggest problem in reducing the cost of maintenance is to assure that any cost reduction in this area does not become an additional cost factor in production. Certainly, it would be folly to reduce maintenance costs and have these savings turn up as additional costs in production from resulting equipment failure." (Behan, Cost Reduction Through Short Interval Scheduling, Prentice Hall, 1972)

Using the strategies of the SIS schedule, we changed some tactics to come up with the Short Interval Response Schedule (SIR). We feel that the SIR schedule is better suited to the maintenance environment.

13.4 Management by Walking Around (MBWA)

The informal approach to this is MBWA (Management by Walking Around). Hopefully you will decide to do some walking around when there is a problem, see the problem, and correct the problem. All management people should get regular direct information from walking around their domains.

13.5 Scheduling Pitfalls

Ron Turley recommends avoiding the following pitfalls:

- **Two People on a Job** — The only time it pays to put two people on a job is when it is absolutely essential. Usually two people will produce the output of 1 ½ when left to their own devices. Be careful with safety assignments (two people assigned for safety). In some cases, one person may injure the other making the job less safe. Look closely at two person crews.
- **Shift Overlap** — This occurs when the afternoon shift supposedly comes in early to "get up to speed." In many shops, the jobs are completed the next day by the same mechanic and in those cases there should be no overlap. In other cases, take a close look at what actually happens at shift change (observe ten or so shift changes). You'll find people going off early to wash up, chat, and read the paper, not update the incoming shift. Of course, the incoming supervisor should overlap and review each job with the outgoing supervisor.

Be sure that the supervisor spends most of their day out on the shop floor. Scheduling works because the greater knowledge and resources of the supervisor can be applied to jobs that are running behind.

13.6 Schedule Set-Up

There are three phases to installation of the system:
- Learning the mechanics
- Understanding the mechanics and the intent
- Complete comprehension, adoption and upgrading

13.7 Installing the System

- Identify and assign unique names/codes to all of the areas where work takes place. These work areas include bays, wash racks, fuel islands, rebuild tables, and the tire shop.
- Design a priority system that will divide ROs into categories.
- Follow through on systematic assignment of labor standards. Use FR, RE, and HS standards, or some combination. Establish milestones for jobs that take more than four hours. Create a book (or database) of the standards.
- Physically build a scheduling board (or two). You can use a 4'x8' sheet of Plexiglas with the permanent information (time, work areas) drawn on a blue print behind it. List the work areas across the top and time down the edge. If 1/2" is equal to 30 minutes then a 48" high schedule will cover eight shifts. Every two hours add a heavy line horizontally. Build a straight edge that can travel up and down the schedule to indicate the current time of day.
- Review all jobs pending and add priority. Add labor standards. Add 10% (or some other allowable lost time factor) to the labor standards for scheduling purposes. This allows some slippage without redoing the physical schedule. The RO has the actual standard on it.
- Load (write down on schedule with grease pencil) jobs starting with highest priority. Actually choose a work area for the repair and a starting time. The elapsed time will depend on the 110% of the labor standard. As you load the schedule, balance the crewing requirements. If you experience frequent emergencies, leave a percentage (equal to your historical emergency hours) of your crew's time unscheduled. Alternatively, leave certain crew members off the schedule when you start-up.

- The schedule should be updated every two hours to visually reflect the actual status of all jobs. Red flags can be used when jobs fall behind.
- If you build two scheduling boards, you can fill the second one with jobs due in the next week. When the next week starts, the board will be up-to-date and ready to go. If you are running on one board, keep erasing jobs completed and start loading future jobs in the section above.
- The two-hour checks should be coordinated with breaks and lunch. In a 7AM to 3:30PM work day, they can be scheduled at 9AM, 11:30AM, and 2PM. During these checks, all jobs are evaluated for status relative to the schedule. Jobs the schedule shows as complete should be complete. Milestones should be met on longer jobs. Shorter jobs that span a check are informally reviewed.
- Numbers are assigned based on completion statistics. A complete job, or milestone met, earns 100%. Jobs not completed on time are evaluated. If an uncompleted job is halfway to the milestone then it earns 50%. All of the percentages are weighted and added. Reports are generated for each supervisor. A supervisor fills out a schedule miss report every time attainment falls below 85%.
- Line supervisor action is expected if the schedule attainment is less than a preset percentage such as 80% or 90%. Frequently, the supervisor can correct these schedule miss conditions. Correction within the next period will bring the schedule on-track but behind. Management decision can change the schedule or one of the floaters (emergency workers) can be used to bring it back up to speed.
- The fleet maintenance manager is expected to intercede only if the daily schedule attainment falls below 80-90%. The fleet manager reviews all of the schedule miss reports. Usually discussions are held with the supervisor, the mechanic, and anyone else involved. If the schedule miss continues for two days or more, the Transportation Manager (or equivalent) gets involved.
- Standard reports with less and less details are routinely fed up to the higher levels in the organization.

13.8 What to Expect

When the system is installed, the response is usually confused at first then extremely positive. Certain people, however, do not want to change. In successful installations, all of the people have been exposed to the system throughout its development.

When there is resistance to the system, it can come from any level. The set-up group should be aware of common reactions and not be caught off guard. Be aware of negative reactions from top management, supervision, and the mechanics/helpers themselves.

13.8.1 Negative Reactions from Top Management
- May stop reacting to and handling problems in their normal manner
- May let installation team handle everything (stay uninvolved)
- May evade responsibility for their operation
- May hide from their people
- May follow system blindly rather than understand it and its shortcomings
- May introduce too many special cases thereby circumventing it
- May raise the objection of too much paperwork for supervision

13.8.2 Negative Reactions from Supervision
- Neglect to follow normal procedures, deadlines, work sequencing or may not react to unusual circumstances as they previously did, stating the system didn't call for it
- Criticize rather than solve problems, unwilling to participate in or support program
- May contribute to grumbling and low morale, voicing doubts or objections to employees to show they're one of them
- Unwillingness to actively supervise, talk to, or discuss problems with employees
- Failure to accept the data as theirs
- Fighting the system, its principles, and basic data rather than doing what the system calls for, taking the least distasteful way out
- Too busy – procrastination rather than direct resistance
- Blaming everything that goes wrong on the system. Using the system as a scapegoat for current problems, as in, "We can't do it anymore. It's not on the schedule."
- Shifting blame to outside departments or vendors

13.8.3 Negative Reactions from Mechanics/Helpers
- Absenteeism suddenly increases
- Employees slow down (as a reaction or on an organized basis)
- Taking an oppositional attitude
- Quality suffers
- Employees walk out
- Petty grievances become major issues
- Sabotage equipment
- Excessive bidding in to and out of jobs makes scheduling difficult
- Employees try to find ways around the system

We are indebted to Alexander Proudfoot Associates for the ideas included in these three lists.

13.9 Work Program

There are two parts to Fleet scheduling that should be in balance:
- The work to be done
- The hours available

Work Program is an exercise that determines the hours available next week. Work Programs are weekly calculations that insure maintenance resources are balanced with maintenance workload.

Work Programs:
- Ensure that expectations for backlog relief are realistic
- Make allowance for all commitments to indirect activity
- Clarify the labor hours required for response to breakdowns, road calls, and urgent jobs
- Clarify the labor hours required to meet PM/PdM requirements
- Define the labor hours of work to be loaded to each crew's weekly schedule
- Ensure that requested completion dates (real or implied by assigned priority) are met
- Ensure that even low priority jobs reach the schedule in a reasonable period of time
- Clarify the capacity to handle all work and thereby determine when to use contract support

Work Programs consist of:

1. Total Available Hours – Gross labor hours authorized including budgeted overtime and contract support. Delete resources committed to various indirect activities including vacation, absenteeism, training, meetings, etc.
2. PM and Routine Work Hours – PM, including PdM, Standing Work (yard services such as fueling, etc.) and Routine Work. The combination of the two consumptions (routine work and PdM/PM) is subtracted from the direct work total.
3. Emergent Hours – Historical amount of emergent, breakdown, and road call work for this time of year or average E demand for last 3 months by week. Average work resources for response to urgent conditions, such as equipment failures causing production downtime, are subtracted.
4. Backlog Relief Hours – The resources available for backlog relief are the hours left after all deductions.

13.10 Case Study – Estimated Weekly Work

SAMPLE WEEKLY WORK PROGRAM WORKSHEET	
Activity	**Hours**
Gross Labor Hours for next week — *40 hours x (10 people)*	400
Indirect Activity Hours for next week — *40 hours vacation + 3 hours safety meeting + 32 hours training*	75
Authorized Overtime/Outside Vendor Hours for next week	20
Total Available Hours for Next Week — *400 hours - 75 hours + 20 hours*	345
PM and Routine Work Hours for Next Week	65
Emergent Hours for Next Week	80
Hours Available for Scheduled Jobs Next Week — *345 hours - 65 hours - 80 hours*	200

All numbers are expressed in labor hours. This is the starting point of all discussions with production about what is possible for the following week.

MRO Inventory   14

If you can get the parts you need immediately, at the lowest cost, without downtime (due to parts) then you don't need to maintain much of a maintenance parts inventory.

One client had a really small stock room for the size of the fleet. It was no more than a broom closet with a few disposables (filters, fuses, etc). When questioned about the seemingly low inventory level, they took the consultant outside and pointed. Across the road was a heavy duty parts distributor! The distributor served as their warehouse (they just had to walk across the street, or run, when traffic was heavy!).

Most firms don't have that geographical advantage. The reality is that usually the speediest source is also the most expensive. Many parts are difficult to get on short notice. Even with a fully equipped stock room, equipment will have downtime due to unavailability of parts.

The function of maintenance inventory is to support the timely performance of maintenance. If you had extra units, unlimited bay space, and no pressing need for any particular unit, then an inventory would not be economically necessary. You COULD wait until NAPA, or the heavy duty parts distributor, arrived with the part.

Another repair facility specialized in exotic automobiles like Rolls Royce, Bentley, Aston Martin and others. The consultant noticed that the shop was quite large for the number of people working. They had a small parts room that stocked only consumable items and a few expensive, unusual spares. When asked about this, the service writer sighed and said parts were the bane of their existence.

He complained, "These OEMs take their own sweet time to ship our spares, often as much as six months. In some cases, they ship some of the spares for the job but not others. The result is that I might have someone's car for nine or more months. Fortunately, anyone that has one of these babies always has other vehicles. Even so, some owners get nasty waiting."

In this case, a proper inventory would be prohibitively expensive. Their customers own vehicles from one to fifty years old, covering potentially hundreds of models.

Owning even an efficient inventory will not (usually) reduce the overall cost of providing maintenance. It will reduce the down time, improve mechanic productivity, reduce the space requirements and make for a faster paced shop. One true savings (if you are careful) is bulk purchasing. Some items can be purchased in bulk which can reduce cost.

Decisions about inventory always lead back to an economic justification that includes the costs and consequences of downtime.

14.1 An Accounting Quirk

In many organizations, when parts are purchased for maintenance, they are directly expensed (unlike raw material inventory for manufacturing which is on the asset side of the ledger, the same side as cash or receivables). As expenses, maintenance inventory is not an asset of the organization. Keep this in mind when discussing maintenance inventory with top management, especially accounting-oriented people. Since the parts are not carried on the books as assets, they are of no use to accounting.

Every accountant knows that one way to increase profit is to cut expenses. In fact, because of this anomaly, if you squeeze the fleet spare parts inventory, profit drips out. For example, when you sell an obsolete part (even below the price you paid), the proceeds are pure profit since the part was charged off as an expense when it came in the door.

14.2 Parts Usage

Modern analysts report that fleet maintenance inventory should turn 4-6 times per year. This means that if your inventory is $20,000, your annual inventory purchases should range from $80,000 to $120,000.

Dividing your inventory items into categories will facilitate analysis to lower your parts costs. Use analysis to uncover the *bad actors* from your inventory. These *bad actors* consume an inordinate amount of the organization's cash, your time, and tie up units while waiting for a part.

14.3 Seasonal Demand

There is a problem with this fine level of analysis for specific types of parts. If you do this analysis on refrigerant or cooling system parts, you might be overstocked for nine months of the year and hand-to-mouth or out of stock the rest of the year. In your analysis, consider the effect of seasons on usage level. Consider this issue when you are designing or purchasing a computer system. A few systems are smart enough to correct the Min/Max and EOQ by the season or time of year.

14.4 Managing the Inventory

Which of the following is the best reason to keep an inventory in your facility?

- **As insurance against downtime.** Having an inventory reduces the need for extra, and standby, units while waiting for parts to come in.
- **A unit that's down is taking up valuable space.** If a repair was already underway, a unit might get stuck in a bay, taking up space indoors. If it's pulled out of the shop, then there is the loss of productivity in moving it in and out, as well as the yard space used.
- **Loss of mechanic productivity.** One loss comes from starting a job, then having to interrupt thought processes, clean up, and move tools to the next job. The loss of productivity is even worse if the mechanic has to jump in a pick-up truck and go get the part themselves.
- **To avoid the cost of having someone expedite the part.** This can be an excessive and frequently unnecessary cost. The cost for air freight can be excessive if the part is not available locally.
- **To avoid cannibalizing good units for parts to repair bad units.** While this is a maintenance no-no, it is sometimes necessary to get a particular vehicle back on the street.
- **To avoid temporary repairs.** Without the right parts, there is a tendency to perform temporary repairs. There is a loss of time in doing an inadequate repair, and how many fleets then forget about the temporary repair and have another breakdown, leading to more lost time.

A Fleet maintenance consultant was sitting on a Boeing 767 being served drinks in first-class during a maintenance hold. According to the first officer, they were waiting for a part.

The Captain came out of the flight deck and began chatting with the flight attendant and first-class passengers. He explained that during his walk around inspection, he noticed a small amount of fuel leaking. He said they were waiting for a short piece of special 767 fuel hose. The problem was the closest part was 300 miles away in their San Francisco depot.

As he sat there, the consultant looked out the window and saw the maintenance people had opened the engine nacelle of the 767 parked next door and were removing the fuel line. They had decided to cannibalize a good plane to fix a broken one. This is a business decision. They figured they would be able to fix the second plane, with the part on its way from the depot, before it was needed in the morning. The problem would have been, had they damaged the part removing it, they would have had two planes down instead of one. Not to mention the lost time to do the second repair. Cannibalization is risky business.

14.5 Calculation of Inventory Carrying Costs

The following chart is based on data compiled from a fleet similar to the one described.

SPRINGFIELD TRUCKING MAIN GARAGE	
Total Inventory Purchases	$191,000
Cost Item	**Amount**
Money (average four turns per year) – 12% for 3 months	$9,100
Warehousing	$4,500
Taxes and Insurance	$3,900
Allocation of People - (25% of purchasing, 25% clerk)	$14,000
Deterioration, Shrinkage, Obsolescence, Cost of Returns	$7600
Total Cost of Carrying Inventory	$39,100
1 + (carrying costs) / (total purchases) = Multiplier for Charge - Out Cost	
1 + ($39,100/$191,000)	1.2047

14.6 The True Cost of Inventory

The charge-out ratio is a multiplier to change from parts price to parts cost. For example, the price for a maintenance-free, 350-amp battery might be $49.50 net. The charge-out cost (entered on the RO part cost column) would be $61.87 (at the Springfield charge-out ratio of 1.25).

When the parts are charged on ROs, using the ratio captures all inventory costs for a more accurate picture of the cost of maintenance.

14.7 Symptoms of Inadequate Inventory Control

The following problems indicate that inventory is not under proper control. Review your operation to see if these symptoms are present.
- Stock-outs on critical parts when they are needed.
- Inventory for units that are no longer in service.
- Inventory for common parts on shelf one year or longer.

- Parts purchased but never used.
- Excessive unusable inventory due to damage or spoilage.
- Inventory cannot be reconciled (receipts plus quantity on hand at the beginning of the period less dispersals should equal new quantity on hand).
- Parts can be added to or taken from inventory without proper paperwork.
- No accountability for parts used or where they were used.
- Purchase orders issued after items received.
- Purchase orders are issued when someone feels that they need a part.
- Everything is verbal, no written records.
- Items are commonly purchased with petty cash or corporate credit cards.
- Little knowledge of parts' location, inventory level, usage, where used.
- Same part stored in multiple locations (and you don't know it).
- No established reorder points, E.O.Q's.
- No competitive bids, sweetheart deals with certain vendors.
- Incoming and outgoing rebuildable spares are jumbled together.
- Outgoing rebuilds are not tracked and managed.
- No attempt to evaluate the quality of the rebuildable vendor.
- No proper storage, unlimited access to parts room.
- No physical inventory taken.
- Hoarding of parts outside of parts room by mechanics.
- No knowledge of the current value of the inventory.
- No analysis of equipment to estimate spare part requirements.
- No knowledge of quantity on hand at the moment.
- Computer's quantity-on-hand is normally wrong.
- Salespeople are allowed unescorted into parts areas.
- Constant calls to vendors for emergency drop-offs.
- Place looks like a tornado hit it.

These symptoms indicate holes, or voids, in your organization's control, management and concern about the inventory. In small fleets, you may have to put up with some of these situations because of inadequate volume to justify elaborate control structures and people. However, even the smallest operations can justify some level of control. Parts are a substantial percentage of the cost of running your fleet. Management will reduce that percentage.

14.8   The Fallacy of Computerization

A computer's ability to do millions of calculations a second is a boon to the inventory field. However, there is a belief that computerization can somehow get any inventory situation under control, once and for all. This is a common misconception. The fact is that without physical and procedural controls (usually entailing a cultural change) computerization will only make the situation worse. The wonderful advantages of computers flow only to organizations committed to controls. Once controls are in place, computerization will greatly simplify the analysis and clerical work.

14.9   The Elements of Inventory Control

These elements include:

- Building adequate storage with limited access, including enough space for all your parts, receiving, incoming and outgoing rebuildables, work area, unpacking and inspection areas, free issue bins (outside the secure area), and an issue counter.
- A storage space that has adequate light in the proper color and appropriate temperature and environmental controls.
- A requirement that all parts removed are recorded on ROs (against RO numbers) or on an equivalent document.
- Parts with assigned locations, and attempts to consolidate parts in multiple locations and stored under multiple part numbers.
- At a minimum, an annual physical inventory taken to verify quantity and location (or cycle counting).
- A requirement that all parts must be received, price checked, physically checked, counted, signed for and shelved.
- Keep parts in boxes they came in with outer box, any plastic, and strapping removed.
- A requirement that all parts put on the shelves have appropriate paperwork and are logged in.
- Some means for recording and tracking usage, price history, where used and substitutions.
- Periodically, parts are shopped, specifications reviewed and vendors evaluated.

- Periodically (every other year), part usage, fleet make-up, and availability are reviewed to adjust reorder point and economical order quantity. Parts are divided into classes for different treatment.
- Parts for units retired and out of service are reviewed for use elsewhere or disposed of.
- The space is cleaned weekly and kept straight.

14.10 The Insurance Policy Inventory Item

Do you have specialized, hard-to-get parts? Normally, if an item in inventory does not get used, you get rid of that part. The Insurance Policy group of parts is a contradiction. You stock these parts for the express purpose of never using them. If your PM system works, these expensive parts gather dust.

These are the toughest parts to discuss with people trained in retail or manufacturing inventories. The rules in those two types of inventories are clear. If it sits, get rid of it. Fleet Maintenance's goal is to provide some capacity (freight hauling, school kids delivered, dirt moved). These insurance policy parts facilitate capacity by lowering downtime (often dramatically) when they are needed.

This class of parts has to be considered differently than other parts. Called insurance policy parts or capital spares, these parts have no analogue in any other inventory system. They have long lead times and are used on high downtime units. These parts can be considered insurance policy parts. They are kept in stock even if they are never used.

Criteria for insurance policy parts:
- Hard to acquire
- Long lead time
- Expensive
- Needed for critical equipment – equipment that has a high consequence for failures.

Consider parts that meet these requirements like your organization considers fire or liability insurance. You buy insurance to manage a risk that you are unwilling or unable to take. You pay the insurance company a premium and they promise to pay when the covered risk occurs. You never want to use insurance but, if you have to, you're glad it's there.

With insurance policy spares, the insurance premium is the cost of ownership of the part. The risk being covered is excessive downtime that the cost or consequence of is far greater than the cost of the premium.

An excellent example concerns stocking of a spare Rotor shaft by a southern utility company. The engine was a large-scale turbine. The Rotor cost a few hundred thousand dollars and had a lead-time of 24 months on a replacement. The cost of downtime for this turbine, during peak load, was almost a hundred thousand dollars per hour (cost of purchasing power from the grid). This insurance policy was well justified.

14.11 Parts Interchange

Truck manufacturing is a global business. Truck manufacturing is actually truck assembly. The components are made by companies located in all parts of the world. One thing to keep in mind is that much of the profit in the truck manufacturing business is in the spare parts business.

Truck manufacturers purchase components (such as axles, transmissions, brakes, wheels, drums, seals and bearings) directly from manufacturers of those items and assemble them onto their brand of truck. The customer specifies all the components for their trucks (or accepts some standard package offered by the dealer). The document that regulates the assembly is the line setting ticket. For example, Fuller transmissions are available from virtually all truck manufacturers.

Here's where it gets interesting. The truck manufacturer (say Navistar) assigns their own part numbers to all the parts inside the Fuller transmission to gain some of the parts business for themselves. Fuller, of course, has already assigned part numbers to their transmission parts.

In fact, most of the manufacturers who use this transmission assign their own part numbers to the Fuller parts. There may be as many as ten different manufacturer numbers for the same exact Fuller part.

There are three reasons why this information can give you an edge in parts purchasing and inventory control:

- **Reduce your inventory**. You may be stocking the same physical part under several manufacturers' part numbers. This is especially true if you have many types of trucks. You may be able to reduce your line items by 10-15% and your inventory level of these items by 5-10%.
- **Reduce parts cost**. Once all of the interchanges have been identified, you will find that different manufacturers charge different amounts for the same part. Each truck manufacturer has a different volume (pays a different price) and has a different mark-up. There may be as much as a 20% difference between the highest and lowest price on the same part. However, your organization may

be purchasing higher quantities of Navistar parts and be getting a better discount than on GM parts. Even if Navistar and GM sold the interchange parts at the same price, your net price would be lower with Navistar (because of the discount).

- **Reduce downtime**. Say you are out of stock of the primary part, for example, MACK part #235414. By searching the interchanges, you might find you have the equivalent Ford part #C6TZ7101A. These are identical to the Fuller part #14326. If you stock-out on all the interchanges, you still have a choice of many vendors, which increases the probability one will have stock. On older or obsolete parts, the interchange route may be the only way to easily locate a part.

14.12 Order Point and Economical Order Quantity

O = Order Point. The order point (also called reorder point or minimum stock level) is where the inventory on hand equals the safety stock plus the expected demand for the item during the lead-time. To get to the order point, you have to calculate safety stock and normal usage first.

$$O = N + S$$

S = Safety Stock. The amount of safety stock equals the difference between the consumption of the item at normal usage levels for the lead-time and the consumption levels at reasonable maximum usage during the lead time.

$$S = L \times (M - U)$$

L = Lead Time. The time it takes to purchase and receive the part, include time from when you discover it has to be ordered to when it will be received (both internal and external lead times).

M = Reasonable Maximum Usage. Determine your maximum average usage during a period of maximum usage.

U = Usage. Certain parts are used at certain rates, add up the usage for a year and divide by the number of months to get the average usage per month.

N = Normal Usage during Lead Time. Determine normal usage for a period equivalent to the time it takes to acquire the part.

$$N = L \times U$$

COA = Cost of Acquisition. Add up the cost of shopping for parts, getting and evaluating bids, issuing a P.O. (Purchase Order), receiving the material, checking the materials and paperwork, shelving the items, approval of the invoice, and issuance of the check. All these costs are called the COA (Cost of Acquisition). Frequently, this number is $30-$100. Many companies have calculated this number as a way to show that acquisition costs are an important part of the equation.

EOQ = Economic Order Quantity. Once the order point is reached, an order for the part is issued and the part is purchased. The amount of the order will equal the EOQ.

14.12.1 Calculating the EOQ

There are several interrelated numbers to determine the critical control numbers for an inventory system. Balancing the cost of acquisition with the cost of ownership gives you the number of parts to order each time. This group of numbers is determined from your history, from your environment or from the part being analyzed.

$$EOQ = \sqrt{(2RS/KC)}$$

R = Annual Requirements. How many of this part do you use per year.

S = Stocking Cost. Determine how much it costs to order any item from a vendor. Stocking cost is also called the COA – Cost of Acquisition. These are the costs of purchasing, receiving, and accounting. For this example, we are assuming $100.

K = Carrying Cost and Downtime Cost. Holding on to the inventory is called COO (Cost of Ownership). You determined the carrying costs in the true cost of inventory section. The downtime cost has to be determined through discussions with the accounting department. For this example, 25% is used for the sum of both numbers.

C = Part Cost. Unit cost of part or material. In this example, we are using $240.

14.12.2 Reducing the Cost of Acquisition (COA)

Although acquisition costs are a smaller part of the entire picture, reducing them contributes to the efficiency of the purchasing department. And even though they don't show up on the maintenance budget, cutting these costs reduces waste and contributes to a lean maintenance operation. The leaner acquisition is, the better the whole operation runs. To cut costs:

- Minimize the number of purchase orders. An easy way to do this is to arrange your purchases so that you have a minimum of $200-$400 per order, but be careful not to hold up buying urgent parts while waiting to fill a purchase order.

- Arrange for your vendor to take a physical inventory, stock your shelves (reducing your labor), remove and replace unused items, and keep the area clean.
- Have your vendor own the inventory (consignment stock).
- Use the Internet and automated systems. Build links between your CMMS and preferred vendors.

14.12.3   Reducing the Cost of Ownership (COO)

The most powerful way to cut the cost of ownership is to own less. The best way to own less is to reduce the usage. Another good way is to buy at a better price. ABC Part Analysis shows you where to look for opportunities for savings. A typical fleet carries a few thousand SKUs (stock keeping units are unique parts). Which ones should you look at to analyze? ABC shows you where to look.

14.13 ABC Part Analysis

A powerful way to analyze inventory is to sort each part by annual dollar volume (price multiplied by yearly usage). In one study, it was determined that the top 7.3% of line items represented 76% of the yearly parts dollar volume. These high moving items that are expensive are called the A items and consume the most money. In the same study, the B and C categories were the remainder of the line items (92.3%) and comprised only 24% of the dollar volume.

14.13.1   Taking Advantage of ABC Part Analysis

- Consider dividing your inventory into A items and other items. You may choose to address some percentage (such as 10%) of the line items as your A items.
- Apply rigorous purchasing and negotiating techniques to A items to lower costs. This is an excellent application for the skills of the purchasing department.
- Review specifications to get better parts at the same costs, or equivalent parts at lower costs. This is where some level of engineering can pay off in big returns. Areas can include lubricants, will-fits, belts, fasteners, and wiring/electrical.

- Apply sophisticated standards to setting reorder point and economic order quantity. Factory production inventory control experts have long studied inventory strategies. The A level items respond best to these techniques.

- Consider creative new vendors, purchasing modes, and approaches. For instance, instead of purchasing fasteners from automotive vendors, go directly to screw jobbers or manufacturers. Another significant savings comes from installing tanks for motor oil and anti-freeze so that you can purchase them by the truckload.

14.14 ABC Case Study

- Usage of two items reduced
- Quality and price of one item increased

A major waste hauler had a problem with excessive use of hydraulic oil. You could almost follow the trucks from the drops of oil left behind. The second problem occurred on some of the routes where the trucks seemed especially susceptible to catastrophic breakdown of the hydraulic system. As you can imagine, digging garbage out of the truck following a breakdown was not popular (especially in the summer after a rain storm).

ABC analysis showed that hydraulic oil was pretty high up on the list. Hydraulic fittings were near the bottom of the A list. A team looked at the statistics, and the trucks themselves, and determined that most of the leaks were from the fittings and the catastrophic losses seemed to be occurring when trees would get tangled up with the hydraulic lines running along the top rail of the body.

The solution was not obvious, but, after discussion and some calls around, it was decided to use a two prong approach. They decided to run a pilot test on ten trucks. For the chronic leaks, they refitted the trucks with aircraft grade fittings. These cost at least four times more per fitting than the automotive grade but seemed not to leak at all in normal usage, and were more robust in general. The second change was to weld some channel iron to the trucks, positioned in such a way as to protect the lines along the top rail.

With some modifications to the position of the channel iron, after a few months it became clear that both solutions were winners. Of course, the cost of the fittings zoomed up as they replaced the fittings on the first ten trucks. But none had to be replaced during the test period. Hydraulic oil usage plummeted, downtime due to hydraulics was down, and no one missed digging out the garbage.

14.15 Another Case for ABC

- Unit cost of one item cut by purchasing in bulk
- Usage and labor reduced while quality improved

ABC PART ANALYSIS

Install tanks for motor oil in a maintenance operation that uses 25,000 gallons per year.

Purchase Type	Cost per Gallon	Cost per Year
Case—24 Quarts	$3.50	$87,500
Drum[1]—55 Gallons	$2.50	$62,500
Bulk[2]—5,000 Gallons	$1.90	$47,500

Savings with Change in Buying Style

	By switching to:	
	Drum[1]—55 Gallons	Bulk[2]—5,000 Gallons
If you currently use:	Your savings will be:	
Cases—24 Quarts	$25,000	$40,000
Drums[1]—55 Gallons	$0	$15,000

1 – Drums can be a cost effective alternative but they are only to be used in organizations with proper handling equipment to avoid injuries to feet, hands and backs.

2 – The current cost of indoor tankage is about $2 a storage gallon. Additional investment in overhead dispensing will dramatically cut waste, contamination, and labor costs on add-oil and oil changes.

There are systems available that attach to oil dispensing systems, automating the chore of feeding information to the CMMS. A data record of unit number, work order number, material (SKU of the specific oil), quantity and time is collected for every transaction. Fluids are a major cost. Their consumption rates (such as engine oil) are leading indicators of problems and are useful to detect problems.

14.16 Using Big Ticket Item Analysis

This type of analysis breaks out big-ticket items for review. In analyzing fleet inventories, a major soft drink bottler found that these big-ticket items (defined by them as $500 and up, per part) consist of 15% of the yearly volume and 70% of the total inventory value on the shelf. Apply this technique to your inventory as follows:

- Use your system to print a list of parts whose unit cost is over some target that you choose. For instance, $500 or $1000. The first time you go through this exercise, pick a higher number to limit the members in the group.

- Determine your 90 day usage for each part. If the lead time is shorter than 90 days, sell or trade the rest of the parts. Note: if the product is difficult to get and the cost of downtime on the unit is high, leave it on the shelf because it is an insurance policy part.
- If you can get the part from a supplier (even at a small premium) consider selling or trading off your entire stock (or letting it run out). Remember that if you can get the part in one or two days, that is equivalent to stock since it usually takes that long to prepare the unit for a major repair (such as an engine replacement).

The premium that you can afford to pay, to get the part from a supplier, is related to your organization's carrying charges and the average amount of time the item spends on the shelf.

CALCULATING THE CARRYING CHARGE OF AN ITEM

An $8,500 engine that normally sits for five months before use in an organization with a 15% per year carrying charge.

Annual Carrying Charge	$8,500 x 15% = $1,275
Monthly Carrying Charge	$1,275/12 months = $106.25 per month
Average Carrying Charge	$106.25 x 5 months = $531.25

You might use 50% – 75% ($266 to $398) of the average carrying charge in determining whether or not to stock the engine. Use 50% – 75% because some of the carrying charges are semi-fixed and would not decrease if you eliminated these items. Another issue on large parts such as engines is that they have a shelf life. The engine could actually go bad waiting to be used.

The resultant reduction to your inventory means you will have:

- Cash available for investments that can pay more significant dividends.
- Your stock room can be better arranged to take advantage of the extra space.
- You increase the number of inventory turns per year, increasing the efficient use of your money.

Purchasing Maintenance Parts

A percentage of your budget dollar is spent on maintenance materials and spare parts. These materials range from tires, to tie wraps, and from batteries to belts. This section is adapted from the work of Michael V. Brown, President of New Standard Institute.

Purchasing must balance cost, delivery, and quality. Material and service purchases account for a substantial percentage of the operating costs in most industries.

The challenge is to find and acquire materials or services at the:
- Right time
- Right place
- Right quantity
- Optimum quality
- Best delivered price

And buyers must be skilled in:
- Negotiating
- Value analysis
- Purchasing law
- Ethics

Why negotiate?
- There is a good chance that any quoted price can be lower.
- Prices will not be lowered until the buyer asks.
- Seller will sell at whatever the market will bear.
- If the buyer is ignorant, the price will be higher.

Remember, when you negotiate, there are only **two** areas in which to negotiate:
- Supplier's Expenses (an option if there aren't many other suppliers):
  ◊ Offer to purchase in larger volume per order.
  ◊ Offer to send company truck.

- Think about the elements that go into the cost of Sales.
- Supplier's Profit (the goal is to obtain a fair price):
  - Supplier will forgo some profit for future business
  - Systems contracts
  - Competition

15.1 The Parts and Materials Buyer

In the U.S. there is a set of Laws of Agency that apply to parts and materials buyers. If you order and receive goods, and your company pays the bill, then you are an agent of your company (in this case a purchasing agent) by law. Agency can be formal or informal and has different levels of authority.

1. If you are appointed by an officer, or have written approval, you have Express Authority:
   - The company is bound to honor all agreements you make, verbal or written.
   - Your authority may be limited by the products and services you may purchase, or the dollar amount you may spend.
   - You are governed by the laws of agency.
   - You are guided by established policies.
2. You informally become an agent when you receive Apparent Authority:
   - Included with other unstated authority:
3. Within written policies and procedures. For example: Ability to authorize air freight.
   - Supplier is led to believe you are an agent:
     - You, or your predecessor, have purchased items before.
   - You have been delegated authority from your superior.
   - You buy things and the company pays the bill.
   - You have been verbally assigned purchasing duties.

Companies should have a policy in place to avoid serious legal problems by giving agency unintentionally. Consider including the following points:
- Only delegated persons can make purchases.
- Buyers have the right to obligate the company.
- Maintain a professional relationship with suppliers.
- Buyers will be courteous, professional and ethical.

- Purchase at the lowest cost and optimum quality.
- Encourage punctual delivery in correct quantity.
- Negotiate return of rejected materials.
- Purchasing manager advises of changes in rules.
- Levels of authority:
    ◊ Perform all purchasing duties
    ◊ Limited to purchasing specific items only
    ◊ Limited to a dollar value per purchase

15.2 General Business Conduct for Buyers

- Give prompt and courteous reception to suppliers and their representatives.
- Treat all suppliers equally.
- Do not take advantage of a seller's error.
- Avoid putting the seller to unnecessary expense or inconvenience on returned goods.
- Keep all specifications and price quotations made by suppliers confidential.

15.3 Ethical Guidelines for Buyers

- Don't lie about your authority.
- Don't take action without authority.
- Never perform an illegal act, even if authorized.
- Don't act against your employer.
- Inform your supervisor of all conflicts.
- Don't have a private interest in conflict with your employer.
- Use your time at work for your employer.
- Don't compete against your employer.
- Never divulge confidential information.
- Turn over all monies.

15.4 Conflicts of Interest

- Do not have outside interests which encroach on time or attention which should be devoted to the affairs of the company.
- Avoid direct or indirect interest in a relationship with an outsider that is inherently unethical.

- Do not take personal advantage of an opportunity that properly belongs to your company.
- Do not use company property without approval.
- Do not buy or sell stock at a time when you have inside information as a result of your position or job within the company.
- Do not disclose company trade secrets, or any other proprietary information, to unauthorized persons.

15.5 Gifts and Gratuities

- Do not accept gifts, personal loans, entertainment, or other special considerations from an individual or business organization doing business with the company.
- Do not accept loans from an individual or organization having prospective dealings with the company unless such individual or organization is in the business of making loans to individuals.
- Do not permit any influence that could conflict with the best interest of the company, or prejudice the company's reputation.
- Buyers should keep themselves free of obligations while attending supplier sponsored luncheons or dinners.
- Any employee who receives an unsolicited payment or gift of more than a nominal value should return it to the supplier.

Vendors

Reduction in unit part costs is one of the ways to reduce your cost of ownership without changing the entire way you do business. Reducing costs means negotiation with your parts vendor to increase your discount. Particular attention should be paid to the items that turnover the most rapidly (the A items).

The USA, Canada and European fleet parts business is very complicated and very competitive, with many vendors involved. As a consequence, there are different discount structures. It is important for you to understand the structures in your area. Large transportation centers have more vendors while small centers will have a simpler vendor structure.

In the fleet field, parts are usually purchased with discounts from the published price list. These discounts have names (such as fleet, jobber, etc.) whose definition and discount varies in different sub-specialties and locations.

The lowest discount is the FLEET discount. Every fleet should be able to purchase at this level. Only the smallest non-stocking fleets should accept this level.

The second tier is DEALER pricing. This is actually only slightly better than fleet pricing. Even moderate sized fleets should negotiate to the next levels.

The discount level is JOBBER level. This is the price small parts houses pay for their parts. Fleets with inventory levels up to $20,000 could save some money at this level.

Major fleets with inventory in the > $20,000 range should be shopping from the WAREHOUSE distributor. WD's purchase directly from manufacturers and usually offer the best discounts. Since they don't carry all lines some flexibility is important.

In some cases, you can purchase directly from the smaller MANUFACTURERS. These prices can be very attractive if you are purchasing in large enough quantities.

16.1 Vendor Selection

When selecting a vendor, if your contract will exceed $50,000 for the next year, and the potential firm is small, then conduct a review of the financial status of the organization. This would include Dun and Bradstreet, trade references, bank references, and the financial situation of the owner (or an annual report if it's a public company). Do they have the financial wherewithal to support you?

Review, in detail, the terms of sale with any vendor. What is their published return policy, stock balancing policy, warranty policy, shipping/delivery/pick-up policy, is there internet ordering, are there any incentive plans (plus any informal deals they have offered your firm).

Ask the vendor for some non-competitive customer references. Ask the reference customers about their experiences with:
- Delivery
- Returns experience
- Stock situation
- Accounting problems and accuracy
- Counter/phone people customer service
- Internet ordering

If you get a chance to visit the vendor:
- Look at their inventory and decide if it matches your needs.
- How does the warehouse look?
- How do the people look?
- Does it look like you'll get the right parts when you order them?
- Do they carry the lines you want?
- If you give them a blanket contract, how will they treat the low volume one-shot type purchases?

16.2 Steps in the Review of Vendors

In order to negotiate with the parts house, warehouse distributor, or manufacturer, you have to know your current prices and the volumes you currently purchase.

1. Establish a parts book (most computerized systems will print this) with current and last price, average price, monthly usage, and usual purchase quantity.
2. Many parts houses respond with lower prices if blanket quantities

are specified, a monthly dollar volume is guaranteed or some other sales assurance is offered.

3. Compare several (level) vendors' net costs against your parts book costs.
4. Pick the best two vendors for each of your fast moving parts. Review their history.
5. Reduce your total list to a few general vendors.
6. By reviewing the prices, swapping parts from list to list, and further negotiation, allocate purchases to each vendor to support whatever guarantees you made to secure advantageous prices.
7. Inform winning vendors then update your parts book or computer system (and generate a new parts book). Inform losing vendors what it took to get the order. Inform second place finisher that they will get the first call if the prime source gets into trouble. Secure next best pricing from second source.

16.3 When to Use Outside Vendors

Outside vendors are an important tool in the fleet manager's pouch. Traditionally they are used for overflow work and other categories including:

- Seasonal (driveway and yard snow removal)
- One time or infrequent work (construction, mounting new truck bodies)
- Low skill work—the vendor can hire at a lower rate than you and has supervision in place to manage these lower paid workers (example: shop cleaning)
- High skill work (transmission rebuilding) or work requiring a license (certified welding).

But the truth is that you can contract out any function, including the complete maintenance of your fleet.

In the best cases, outside vendors and contractors are used when you:

- Want to save money
- Want to improve quality—specialists perform the same work every day so they get quite good at it
- Lack a skill—transmission work, spring work, etc.
- Lack an appropriate license or skill set—such as frame welding or tank fabrication

- Lack specialized equipment—frame straightener
- Want to reduce your liability—elevators, fire systems
- Want to reduce the hazard to your own employees—window washing, asbestos removal.
- Want an outside opinion (a devil's advocate or skilled tradesperson or engineer) or you need an outside expert to show you a whole new approach
- Need training—have your mechanic help, or tag along, to improve their skills
- Are already busy or the job is too large
- Don't want to manage a job—hire a contractor
- Want flexibility
- Run into politics—you disagree with top management about the number of hours a job should take or other political reasons
- Don't want to lose control of existing projects to make room for a large new one.

Use outside vendors in rebuilding and projects when:
- You lack internal expertise or your internal people are not available.
- Work is intermittent (not long term) or is used in peak demand shaving.
- Cost internally is higher than externally.
- External quality would be higher.
- Tools required to set-up are too expensive.
- No space is available.
- Span of control would get too wide to take on job internally.

Outside vendors should not be used when:
- Internal expertise is available.
- Project is of a continuing and long-term nature.
- Project is not well-defined.
- Cost is significantly lower internally.
- Expensive internal equipment is available.

Specific Criterion for selecting a major component rebuilder:
- Does the rebuilder offer a reciprocal warranty with rebuilders in other locations?

- In addition to the rebuild, is the component updated to the latest engineering revision of the manufacturer? Is the component rebuilt to OEM specifications?
- Can the rebuilder give you an analysis of why the component failed and keep records on previous rebuilds of the same component?
- Will the rebuilder stock your requirements on a consignment basis?

Warranties 17

How much warranty is left on the table? In other words, how many fleets collect all the warranty money that they are entitled to? Manufacturers tell us that they reserve 1%–2% of the sales price of a truck for claims. Most fleets do not collect the full amount of warranty that they are entitled to. In short, a great deal of money is available to all fleets and only a fraction of it is collected.

Warranty is money owed to you from the manufacturer of your trucks, trailers and other units. This warranty recovery discussion also includes large tools, and building components such as heaters, roofs, lighting, in short everything you buy! As such it should be managed as intensively as any other receivable.

In an article in the July 2004 Maintenance Technology magazine, Joe Mikes discusses a warranty recovery system. While he was not referring directly to vehicle warranty, most of his points can be adapted to the fleet field. He stresses the importance of reading the warranty to be familiar with the contents and negotiate some additional clauses if you are a large enough buyer.

One of the issues with warranties is that, to be effective, several departments have to work together. In smaller organizations, a single person might be responsible for all functions.

- Purchasing has to negotiate improvements/clarifications to the existing warranty.
- CMMS has the rules for the warranty and issues an alert if repairs are being done during the warranty period.
- Maintenance management initiates the claim.
- Data about the parts are collected from the store room.
- Accounting, or accounts receivable, manages the incoming money.

17.1 Warranties on New Equipment
- What is the length of the warranty period expressed in either distance traveled, hours, or days?
- What can disqualify the warranty?

- If something breaks does the warranty start over?
- What PMs are required and what proof is acceptable?
- How long can a dispute go on?
- Is there a defined response time from the vendor?
- Is there a lemon clause (if something breaks over and over they replace it)?
- What data is needed for a valid claim?
- Who pays shipping and airfreight if the dealer is out of the part?
- How is warranty work conducted and what documentation is needed?
- Do you have to be approved to provide your own warranty work (to reduce the cost of shuttling vehicles back and forth)?
- What is the warranty labor rate?
- What standard is used for labor?
- What about replacement units if a unit is out of service for an excessive time period?
- What about reimbursement for shipment loss or spoilage (for time sensitive loads)?
- If equipment is replacement, can you upgrade it?

17.2 Parts Warranties

While most fleets recover some money for warranty from new vehicles, there are fleets that recover no warranty money from defective parts.

Joe Mikes (in the same July 2004 article) outlines the information you need to collect for any claim:

- The unique ID number on the part that failed.
- The date and location where the part was bought.
- The date and/or meter reading when the part was put into service.
- The proof of purchase.
- The warranty guidelines for that part (6000KM, 3 months, etc.)
- If the part is also a major asset such as an engine, proof that PM was done (such as oil changes).
- Where there is a decent amount of money involved, photographs of the broken part might be useful.

Where does all this data come from? The primary source of data is the work order system:

- The work order has the unique ID number of the unit.
- A different work order has the actual date and meter reading when the part was put into service.
- The warranty guidelines should be filed in the part master file that has all the fixed information for each part.
- The purchasing department has proof of purchase but the reference number (purchase order number) is in the part transaction record.
- To reconstruct PMs for that unit, query the work order system for all PMs between date in service (of the part) and date of breakdown (again of the part) for the unique unit number.

A digital photograph is a useful addition to the documentation package for expensive parts or where parts failed in unusual ways.

This package is sent to the vendor with an invoice for the labor. Using the manufacturer's labor guide for labor hours helps reduce the number of topics in the discussion. Usually a replacement part is adequate reimbursement for the failed part. Of course, there will be a large discussion about the labor charge.

17.3 Reusing Parts

The general recommendation is that fasteners should not be reused. The fastener works by clamping the joint. The clamping force is generated from a slight stretching of the bolt. When a bolt is in service it might change characteristics from the stretching. That being said, most shops do reuse bolts. Do not re-use bolts in critical safety applications.

17.4 Cost of the Failure of Failure Analysis

Fact—Electrical part manufacturers report that 75% of all parts returned are operational.

Fact—Ron Turley, a fleet maintenance consultant, reported on a fleet that was spending $300,000 a year over-lubing their fleet. He reports that a six week cycle would have been adequate but they were lubing all units on a weekly basis. The reason for the frequency was a mistaken failure analysis of a repetitive U joint failure. They had decided that the U joints were failing from lack of lubrication when, in fact, they were being repaired incorrectly. No amount of lube would have reduced the failure rate.

Fact—With the new "brains" that control vehicles, the main method of repair is to replace things until the unit starts to work again.

Fundamental Questions About Fuel

18

You should know the answers to these questions!
- What are my total fuel purchases?
- Where do I purchase my fuel?
- What are the different prices by source?
- How much do I use?
- How much is missing?
- Why don't I fuel in-house?
- What is my fuel specification?

RELATIONSHIP OF FUEL SAVED TO MONEY SAVED

100 unit fleet @ 6 MPG, averaging 100,000 miles /year/unit
Uses 1,666,667 gallons/year and spends $6,666,668 on fuel at $4/gallon

Savings	100,000	250,000	500,000	1,000,000
10%	$40,000	$100,000	$200,000	$400,000
9%	$36,000	$90,000	$180,000	$360,000
8%	$32,000	$80,000	$160,000	$320,000
7%	$28,000	$70,000	$140,000	$280,000
6%	$24,000	$60,000	$120,000	$240,000
5%	$20,000	$50,000	$100,000	$200,000
4%	$16,000	$40,000	$80,000	$160,000
3%	$12,000	$30,000	$60,000	$120,000
2%	$8,000	$20,000	$40,000	$80,000
1%	$4,000	$10,000	$20,000	$40,000

(Gallons)

18.1 Fuel and Time

Another issue is time. How long does it take to fuel? Modern pumps can peak at 19 gallons per minute (or they froth and you have to wait for the fuel to settle). In the previous example, this fleet is filling its tanks 13,333 times at 125 gallons per average fill.

The most efficient in-house fuel set-ups have high speed pumps and a master-slave fuel pump (allowing both saddle tanks to be fueled at once).

18.2 Fuel Consumption

After the oil embargo in 1973, and again in 1979, fuel efficiency became a national priority. There was a great push in the U.S., in the 1970s under President Jimmy Carter, to improve efficiency. The CAFÉ standards were adopted and significant research into fuel efficiency began.

Much of the data for this section is derived from materials generated from the U.S. government's Voluntary Truck and Bus Fuel Economy Improvement Program (VTBFEIP). This 1970s program conducted research and disseminated information from manufacturers, users, and associated parties. All parties involved published their own studies on the influences of fuel consumption and the components of a fuel savings program. Unfortunately, this valuable source was cut from the federal budget and has been discontinued. For three decades, relatively low fuel prices and lack of interest have slowed progress toward increased fuel efficiency. Well, the pendulum has swung back. Due, not to a single crisis, but rather, to a series of crisis throughout the oil producing world that has resulted in peaking fuel prices.

The VTBFEIP, and its members, commissioned excellent research into the factors of fuel consumption and thoroughly investigated fuel savings techniques and devices. Following is a review of the factors in fuel consumption.

18.3 Factors in Fuel Consumption

Fuel is consumed to generate power to turn the engine. Power is needed for four purposes:

Inertial Resistance—It takes power to overcome inertial resistance to start the unit moving or to accelerate the vehicle to needed speed. Power to overcome inertial resistance is determined by acceleration and gross mass. In addition to standard inertial resistance there is grade resistance. The amount of power required for long constant grades is:

$$HP\ Grade = 0.03645 \times GM \times G \times V$$

where:
- GM is gross mass
- G is grade in %
- V is velocity

Note: This is a metric formula, expressed in Metric Tons, KM/HR.

Friction—Friction causes rolling resistance. Rolling resistance is the tendency for the unit to slow down unless engine power is applied. It is the internal resistance of the bearings, drive-train friction, and the viscoelastic resistance of the tires against the pavement. The vehicle mass, road surface, and design all impact rolling resistance.

Aerodynamic Drag—As the unit moves faster, aerodynamic drag resistance (wind resistance) becomes a big factor. At 63 MPH, wind resistance consumes 50% of the available power. The standard square shape of a truck (contrast truck shapes to airplanes or speed boats) requires high amounts of fuel to merely part the air. The height, width, shape and speed determine this resistance. Doubling the height or width (for the same shape) will double the wind resistance. Speed of the vehicle (or a head wind) has a major effect. Doubling the speed will increase the drag by 8. Of course, better tractor shapes reduce the drag but may be a problem for combinations pushing the length limits.

Accessories—including air conditioning, thermostatic fans, alternator, air compressor, power steering pump, etc. These items' use of power varies with engine speed (rather than road speed). The total of these loads varies from 3–4HP under light conditions to a maximum of about 35HP.

Total Power—The total of the four factors is the required power to do the job. The margin between the power needed and the power available is apparent for hills and acceleration. Note that the margin decreases as speed increases. Items that will impact fuel consumption will impact power. The sum of these four factors is the minimum total power requirement.

18.4 Fuel Saving Recommendations

The VTBFEIP and its members recommended areas to reduce fuel consumption. It's interesting to read the list and see how many of their ideas are part of fleet operation today and how many are still an issue, forty years later.

- Steel belted radials reduce rolling resistance by 40% over bias ply tires.
- Super singles decrease resistance additionally as well as provide mass (weight) improvements. The quality of the rubber compounds, and mileage life, of these tires has significantly improved recently. If you haven't tested them recently, they might be worth a look. Fuel savings of 0.2 gallon per mile are reported. There is also a savings in maintenance costs since there is only one easy-to-find stem.

- Make sure tires are properly inflated. Proper balance is also important in reducing the amount of work the tire does, as well as increasing tire life.
- Improved lubricants can decrease rolling resistance 2–3%. These include ester-based lubes or ones with additives, such as molybdenum, Teflon, or graphite. Additives have been shown to improve efficiency and reduce wear and tear. Run your own tests to separate the flashy ones from the ones that work (only by testing your fleet in your environment will you learn which ones live up to their claims).
- Purchase smooth sided trailers with well-rounded corners. Fuel consumption gains with farings are well established but knowledge of their proper use is important. Roof farings are only effective when used with vans. Fuel performance drops when farings are used on dump trailers, flats, and bobtails. To maximize fuel efficiency, the trucks/tractors of the future sport smooth wheel shirts, panels between the tractor and trailer and smooth curves on all corners. Smoothing and Rounding will increase the efficiency of straight trucks and busses, too.
- Improve operational techniques. Under-fill fuel tanks in hot weather (for expansion see following). If possible, garage vehicles in cold weather to cut fuel on start-up. Improve routing to reduce mileage. If you are cube limited, doubles will provide better fuel gallons per ton-mile. Running with full loads also improves efficiency and $/ton-mile.
- Thermostatically controlled fans will reduce fuel use. In tests on 23 units, the fans ran less than 3% of total engine hours. This can save 20HP on large engines. See following discussion on APU.
- Re-specify engine to more nearly match HP actually required, including turbocharged and reduced fuel consumption engines. Consider re-powering before you are about to do any major work on the engine. Specify gear train to minimize RPM (hit the RPM sweet spot for your engine choice) at road speed.
- In all cases, lightened components will reduce fuel consumption. Consider aluminum bodies to cut weight (and increase life).
- Govern engines to limit maximum speed. Remember, wind resistance increases eight times as speed increases twice. Also, govern engines to operate at maximum fuel efficiency which is usually near the maximum torque RPM.
- Keep vehicles maintained. For example, avoid a smoking diesel, keep air cleaners free, use a torque wrench and don't estimate

torque of heads, keep wheels and axles aligned (inspect frame for alignment), keep tire sizes matched, keep tire pressure right, and adjust brakes (so they don't drag).

- When replacing components, choose those that will increase efficiency. Key areas include replacing mufflers using low back pressure units (check for dents or twists that would increase back pressure) and changing to disk wheels.

- Improved driving practices can save 1–10% of your fuel bill. These include observing speed limits, keeping RPMs down, maintaining steady road speed, gradual acceleration, shutting off engine when not in use, avoiding overfills, and shifting as little as possible.

- Watch the fuel you buy. There are many ways fuel can disappear. It can be pilfered from vehicles, spilled, pilfered from tankage, never received, etc. One trick is to heat the fuel so it expands in the tank truck prior to delivery. Once in the ground, it cools and shrinks. At service stations in hot climates, they buy 60 degree fuel (dense) and then dispense it at ambient (less dense) temperature. There has been legislation on this scam. The delivery meter should be temperature compensated (known as VCF—where the volume is adjusted to 60 degrees). This is why you don't top off the tanks in warm weather.

If we add up all of the ways fuel dollars are consumed, the potential for savings is dramatic. Remember that fuel consumes about 20% of total fleet dollars.

18.5 Biodiesel

One of the biggest movements in the field is biodiesel. It extends from diesel made from canola oil to converted French fryer oil. Once, biodiesel was a far out alternative fuel now it is starting to move into the mainstream. Europe is ahead of the U.S. in this area. Biodiesel has some advantages for the country. The biggest argument for biodiesel (aside from national security) is that diesel production is tied into communities. The producers can be local plants running local feed stocks. This community orientation is attractive to many fleets.

Almost any diesel can run on low percentages of biodiesel. Most engines today can be adapted to run on more concentrated biodiesels such as B20 (20% biodiesel). Some people claim to run 100% biodiesel with no ill effects. There is good evidence that biodiesel is more slippery than petroleum based diesel resulting in better lubrication and longer engine life.

Mark O'Connell, editor of Fleet Magazine, wrote in a 2008 editorial that there were many misconceptions held by fleets about biodiesel. He cited lack of quality standards, lack of filtering, and rotting of aluminum fuel tanks as popular misconceptions. This industry is developing and clearly needs to conduct outreach and educate fleet managers.

More research is clearly needed (as of 2010). One of the active groups, Sustainable Biodiesel Alliance, is seeking fleets that are interested in biodiesel but have concerns (www.sustainablediesealalliance.com).

18.6   Fuel Efficiency Gadgets

Selling fuel efficiency gadgets is a popular business. From cow magnets to devices that inject small amounts of water, where there is money to be made someone will try. A smart fleet manager will test these gadgets under controlled conditions before being satisfied that the devices really work.

Mark O'Connell, editor of Fleet Maintenance magazine, writes in his August 2004 issue about a savvy fleet manager who was offered tires that would improve fuel efficiency. The manager installed a set of the new tires on one rig and a new set of his standard tire on another unit. Both rigs had identical typical loads. In the test, both rigs were accelerated to the speed limit and allowed to coast. The rig equipped with the new type of tire rolled 500' further than the one with his standard tires. Since the fleet manager had control of all other variables, he could be assured the tires were the variable that improved performance. Of course, a fleet manager/scientist would switch the tires between the rigs and re-run the test to be sure his results were accurate.

18.7   Idling

California's five minute idling rule—CARB (California Air Resource Board) is a game changer in both hot and cold climates. The regulation (soon to be duplicated in other states) is practically insuring that APU's, electric hook-ups, and A/C-Heat ducts to the cabs become more common.

There are two major classes of solutions to the dilemma within the vehicle. The APU (Aux Power unit or Alternative Power Unit) and the battery based system. Of course, within the two classes there are a myriad of alternative approaches.

One typical gadget that could save money is the APU (aux power unit) coupled with accessories that are powered by electricity. The APU (small diesel generator) can generate enough electricity for A/C, radio, interior lighting, and heating at about 0.2 gallon per hour (as compared to 1.0–1.25

PPH for the engine). An average truck in the U.S. consumes 1% of its fuel idling. The number climbs considerably in the South in the summer and in the North in the winter. Several states in the U.S. have new programs to install electrical plugs at truck stops (called the Truck Stop Electrification project) so that electrical accessories can run without the APU, saving even more fuel.

The battery systems have extra batteries and 12V A/C units. They are designed to isolate the battery system for comfort, from the cranking system. In cold weather, they sport supplemental diesel fired heaters.

18.8 Computerized Fuel Management Systems (FMS)

There are five reasons why organizations invest in Fuel Control Systems (Fuel Management Systems):

1. Security – Only authorized people can get fuel. The systems are driven by cards, scanned keys, tokens or codes and only insiders have access. Lost codes or keys can be locked out of the system.
2. Accountability – All fuel pumped is recorded against an authorized account. Individuals using excessive fuel can be detected.
3. 24 Hour Availability – Your fuel is protected around the clock. No need to lock the area for fuel security.
4. Saves Time – No looking for the padlock, no writing down the transactions, no chasing missing log sheets, no late calls for people who have to fuel up.
5. Automatic Record Keeping – Permanent records of transactions for auditing, fuel efficiency and maintenance usage are created. No extra effort is required to accumulate this data.

One of the best stories about the effectiveness of electronic Fuel Management Systems took place early in their development. One of the pioneers in the electronic FMS field was a UK company, Bowen Automation. Their BA series fuel recorders were masses of relays and transistors, quite primitive and unreliable by today's standards. All of the machines needed regular service to keep running. A municipal garage north of London, was their second customer.

This second machine ran without a callback from the customer (this was unusual). When the customer was called, he reported that everything was fine. Two years later, one of Bowen Automation's salespeople decided to use the site for a demonstration to a new potential customer.

Everyone was surprised to learn that the machine didn't work. In fact, the machine broke down a week after installation. When questioned, the superintendent said that when the system was installed, fuel consumption dropped 10% and it was still down. The drivers didn't know the system was broken and he didn't tell them. He was satisfied.

One of the issues in the fuel management industry was its reluctance to identify itself as a small part of a larger fleet information system. These vendors imagined that they (literally) drove the fleet. Recent developments in technology have made it easier to be part of the overall organization.

18.8.1   How FMS Operates

- There is a card, key, token, or code introduced to the system. The system checks to see if the token (security device) has been authorized on the site.
- After authorization, the system may prompt the driver for odometer reading and/or PIN number. Usually input of information is similar to a hand calculator with a clear key and an enter key.
- The system is then ready for pump/product selection. Frequently this information is coded on the token or entered into the system. After selection, the pump power is turned on.
- The amount pumped is constantly being updated until the transaction is complete. Transactions are completed by removing the token or by timing out on the last fuel pumped.
- After completion, the transaction is usually printed, summarized, and stored.

18.8.2   Retrofitting or Installing a FMS

A device is mounted in the register head (the part of your gas pump that displays the gallons) to indicate every 1/10-gallon. This device is called a pulse transmitter or pulser for short.

The electric power is rerouted through the control section of the FMS. The FMS directly controls the electricity to the pump or can initiate a reset signal to restart the pump. This control provides the security function. In some cases, it is not necessary to reroute the power, all that has to be controlled is the reset motor or reset circuit.

A stand is mounted near the pump to enclose the reader/ keypad/ display portion of the FMS. A console may be placed indoors to control printing, lockout, and other functions.

In the U.S., all of the installation changes have to conform to the explosion-proof standards of the locality (usually based on article 500 of the National Electrical Code).

18.8.3 Problems and Solutions with a FMS

- These are usually high techs devices mounted outdoors that need periodic attention. Have an inside person, trained by the installer or factory, PM the unit. Add to your PM list.

- Like other high tech devices, FMSs are sensitive to problems in the power line. Occasionally, fuel pumps or electric reset brush motors will cause problems. It may be necessary to isolate or filter noisy motors from the FMS.

- Sags, dips, brownouts, noise, surge, and blackouts can be problematical for FMSs. Some systems are completely protected and some systems have minimal protection. Some units require a clean electrical line that runs directly back to the breaker panel. Check and verify nothing else shares that circuit. If you experience problems, think of the FMS as a computer (since it is one). The same isolation, filtering, and protection devices will work. Check the load first since the FMS load usually includes the pumps!

- In some cases, the failure of the gas pump motor can create a problem for the FMS. As certain motors fail, they use high levels of starting current which can pull down the line enough to shut down the FMS. Improvements to the power supply in the FMS usually help. Keeping your pumps in good shape will add to their life as well as the life of the FMS.

- Don't mount the communication or power wires overhead unless they are optical (fiber). Overhead wires can pick up lightning that can destroy the electronics. If possible, bury all wires going to and coming from the FMS.

- Most systems have weather-tight cases. Be sure to position the unit away from the wind so that dirt and water can't blow in as easily. While most systems do fine in the weather, all systems do better under shelter.

- Position reader so that a passing vehicle cannot hit it.

- Position the inside console so that no coffee or soda will spill on it. An extra cover wouldn't hurt either (keep the vents free).

- The most common causes for service are blown fuses, out of fuel, failure of the gas pump, and dirt in the token reader (card reader, key reader, etc.). Require drawings of the power wiring and maintenance procedures materials. Plan to take care of this level of service.

Tire Management

Tires are your highest operating expense after fuel. In a Tractor/Trailer combination each mile wears 18 tires (18 tire-miles). With the cost of new radials in the $300 to $400 range, tire costs must be controlled.

Today, you must factor in purchase cost, retread costs, number of safe caps, tire mileage per 1/32, theft, cost of carrying inventory, PM costs, costs of managing the rolling asset and warranty recovery (policy adjustments).

19.1 Elements of Tire Management

- Brand tires to track them individually.
- Scrap Analysis—Track why items are failing. Issue failure reports by reason, age, make, model, etc.
- Manage and track recapping, if it pays in your environment.
- Choose recapper on performance. Review for new technology.
- Become knowledgeable about new tire specifications. Purchase the right tires in the beginning. Specify tread for maximum life in your application and for highest fuel efficiency.
- Don't give good tires away on traded equipment.
- Accountability—Know where your tires are and control theft, swapping, retreaders who skim, invalid road call expenses, normal loss and misplacement.
- Track performance by make and model of tire. Know the best tire for your application.

19.2 Tire Management Systems

There have been several good attempts at computerized tire management systems. So far, all of the attempts have failed to gain traction (sorry) in the marketplace. The tire tracking portions of the larger maintenance management systems tend to be used like fancy inventory systems rather than tire management systems.

The software design for a good tire management system is not difficult. The problem has been a physical one. The magnitude of the data collection does not currently show a good return on investment. The antidote is in technology.

How do you record and track the utilization on (up to) 18 tires by position? Every mile the unit logs has to be added to 18 tire files. Every tire move and repair has to be tracked by both the unit number and the tire number. You can see the magnitude of the problem especially when you include road repairs, purchases, shop repairs/swaps and theft. A bulk of the problem can be managed with chips that can be embedded into the tire rubber.

A tire management system will automate the paper functions of the elements of tire management from the previous section.

19.3 Tire Pressure Systems

According to NHTSA (National Highway Transportation Safety Administration) 500 people are killed every year, in the US alone, from tire defects. The NHTSA has published that, as of the 2008 model year, all cars and light trucks have a pressure monitoring system. This is the latest federal act starting with the TREAD act of 2000 (that was written after the Firestone/Ford Explorer tragedies).

Scott De Laruelle reports, in an article in Fleet Maintenance magazine (March 2008), that these devices measure tire pressure and set off an alarm when the pressure falls 25% below the recommended levels. One could argue that this technology is needed in cars and light trucks but not in fleets run by professionals. Yet the FMCSA's (Federal Motor Carrier Safety Administration) research shows that less than half of the commercial vehicles are within 5 PSI of target pressure, and 7% are under inflated by 20% or more.

With the impetus of the NHTSA regulation, vendors have had a renaissance in sales that has lowered prices and expanded capabilities. Short range wireless transmission means that the tire monitors do not have to be wired to the cab. A simple receiver can do the trick.

19.4 The Best Tires for the Job

Choosing the best tire might include tradeoffs: high cost, low rolling resistance, long mileage, or strong sidewalls. The advantage is that today, some tires have several attributes. The most important thing when you change tire purchase specs is to (as scientifically as possible) test the tires in your main types of service. Some of the minor specs might make a huge difference.

There are several arguments regarding the best tires for the job. For most high mileage, over the road applications the tubeless radial with recapping is the most economical choice. Other situations indicate other combinations.

COMPARING TIRES	
Radial	Bias
• 40% longer life • 5% increase in MPG • Increased puncture resistance • Improved traction • Higher load rating	• 35% less expensive • Increased sidewall strength
Tube	Tubeless
• Nails tend to stay in tire • More repair for retreading • More damage to casings • Less down time	• Quicker to mount • Less storage • Lower cost • Less weight

We are indebted to Joe Pearce for his work on tire management techniques and his presentations at the ATA that provide a framework for viewing tire costs. Some material was also adapted from the Pepsi Cola Management Institute—Fleet Management Seminar.

19.5 Practices to Lengthen Tire Life

This section was partially adapted from an article in the Good Year magazine by Al Cohn, Marketing Manager for Commercial Systems Engineering.

- Follow proper installation procedures. Always use a new tube for a new tire. Never over-inflate a smaller tube to fit a larger tire.
- Ensure there is enough space between duals for adequate airflow to cool the tires. Match duals by type and tread depth. Measure all four tires on the axle. They shouldn't vary by more than 1/2" in diameter. If they are within 1/2", mount the smaller one inside.
- Make sure load ratings are not exceeded—a 20% overload = 30% less tire life.
- For proper positioning of casings, start with steering axles, then drive axles, then recapping, moving back to trailer axles. Specifically, "If you own your trailers, we suggest you remove steer tires at 6/32s to 8/32s and run them on trailer axles down to 4/32nds. Then pull these for retreading and run first retreads as drive tires and second retreads as trailer tires. Meanwhile, you should remove your drive tires by 4/32nds if your trucks are in line haul service and use first retreads as drive tires and second retreads as trailer tires" by Al Cohn.

- Proper inflation for tire type, service and load is very important. Under-inflation causes excessive heat build-up. The extra flexing that causes the heat also increases fuel consumption—10% under inflation = 5% loss in fuel performance. Over-inflation causes the center of the tread to wear faster. Excessive pressure causes abnormal rubber growth, which can stretch and fail. Always check tires when they are cold from not running, and make sure you catch the inside duals as well as the outside duals. Use a calibrated pressure gauge, but remember that your gauge can lose its calibration every time you drop it.

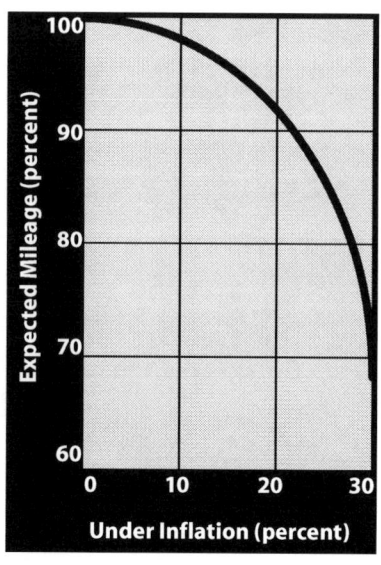

- Reduce abuse caused by oversight, which can reduce tire life expectancy 5–10%, such as buffing usable tread, running mismatched tires, improper scrapping procedures, and poor scrap analysis.

- Maintain drive mechanisms to improve tire life. The Maintenance Council (TMC) recommends that a new vehicle be checked, after going into service, in 90 days or after 15,000 to 30,000 miles. TMC alignment means both toe and drive tire thrust angles plus camber. And don't overlook trailer alignment. Even if your tractors are in perfect alignment, out of alignment trailers are going sideways down the road and hurting the performance of your tractor tires.

- Driving habits can contribute to long tire life. Common sense things like slowing down for potholes, and avoiding driving over curbs and on shoulders, contribute to longer tire life. As mentioned earlier, heat is the enemy of tire life. Increasing highway speeds from 55 to 75 mph (which will increase heat), can reduce your tires' life by more than 30 percent.

- Inspect all tires for road hazard damage, sidewall snags and cuts, sidewall ozone cracking and irregular wear. Also see that valve caps are on tight, or use flow-through air valves. Prompt repair of all damage can save money.

- There is good research showing that filling tires with Nitrogen can save money. In an article in Fleet Maintenance magazine (May 2008), Beth Grahn notes that research shows that Nitrogen filled tires lose air (Nitrogen) more slowly and run cooler. 400,000,000 miles of tests have been run by NASA, Boeing, US Military and NASCAR. The tests showed a 40% increase in tire life and a 6% increase in mileage (if you do not have an effective tire management program) with the use of Nitrogen. The fuel savings were less (2.8%) with an effective monthly tire checking program.

19.6 Case Study on the Costs of Buffing

Buffing is the removal of a tire before it is necessary.

MINIMUM ALLOWABLE TREAD DEPTH BY US LAW	
Steering axles	4/32"
Drive axles	2/32"
Trailer axles	2/32"
Annual Cost of Using Full Allowable Tread vs. Buffing	
Fleet of 100 trailers—each unit travels 75,000 miles/year	
Average trailer tire tread depth from retreader – 6/32"	
Assume a cost of retread of $150 less $35 for case (at end) gives you $115 for 14/32" of usable tread	
Assume 10,000 miles per 1/32"	
If tires are removed at 5/32" (rather than the allowable 2/32") then 3/23" usable tread (or 21% of the capacity) is lost.	

Total Tire-Miles per Year	75,000 miles per year x 100 trailers x 8 tires per trailer	60,000,000
Total Tread Consumed (in 1/32) per Year	$\dfrac{60{,}000{,}000 \text{ tire-miles}}{10{,}000 \text{ miles per } 1/32"}$	6000
Average Cost (per 1/32) of Tread by Using Full Allowable Tread	$\dfrac{\$115 \text{ per tire}}{14 \text{ usable } 32^{nds}}$	$8.21
Average Cost (per 1/32) of Tread when Buffing 3/32 of Usable Tread	$\dfrac{\$115 \text{ per tire}}{11 \text{ usable } 32^{nds}}$	$10.45
Annual Cost Using Full Allowable Tread	6000 x $8.21	$49,260
Annual Cost with 3/32 Buffing	6000 x $10.45	$62,700
Yearly Savings with Use of Full Allowable Tread	$62,700 - $49,260	$13,440

Other considerations of buffing usable tread:
- The last few 32nds of a worn tire actually last longer than the first few 32nds.
- The MPG increases as the tire wears—with less rubber there is less rolling resistance.

Leasing Mobile Equipment

One of the most popular alternatives to fleet management is leasing (full service and financial). Many excellent companies provide this service. Full service lessors range from the billion dollar giants, like Ryder and Leaseway, to small local firms who lease trucks to themselves, and, occasionally, to outside companies.

In legalese, a lease is a contract conferring a right on one person (called a lessee) to possess and use property belonging to another person (called a lessor). Everything else is negotiable.

The words Lease and Rent have similar legal meanings. Generally, a lease is a long term contract and a rental agreement is a short term contract. The rights and privileges of both parties can be similar.

There are many possible reasons for leasing (renting) instead of buying. For example:

- Rental (or lease) payments for equipment used in a business are completely tax deductible as an expense (unlike acquisitions which are depreciated).
- Renting keeps the debt off the balance sheet, unlike purchasing an asset.
- Lease payments are usually lower than payments on a loan (because the residual value of the asset does not have to be amortized or paid back), and qualifying is usually easier.
- A company might want to keep cash in the bank or credit lines open for other business purposes.
- Some businesses prefer leasing as it allows them to return a unit and select a new one when the lease expires.
- There is no need to worry about lifespan and maintenance.
- A lessee does not have to worry about the future value of a vehicle.
- The renter can leave the burden of upkeep of the property (PM, maintenance, repair, body work, etc.) to the owner or his agents.
- Leasing reduces the financial risk, especially during peaks which might only last a short time.

- Leasing is advantageous when something is needed temporarily, as in the case of a special tool or truck.

The full service lessor provides maintenance, insurance, taxes and tags, along with the vehicle. Leasing (or rental) can also be "wet" or "dry". A wet lease includes fuel and an operator (a common way to rent a crane). A dry lease includes the vehicle, maintenance and insurance. This can be an excellent alternative to ownership. Let's look at the costs of a typical leasing operation. As we do this, you will uncover opportunities for self-management.

20.1 Ownership Costs

The leasing company purchases equipment every day and have experience in negotiating favorable purchase prices and a familiarity with the local markets. A smart buyer can get similar deals if they are willing to shop and take time to become experts in their specific equipment markets.

Leasing companies usually borrow significant amounts of money. Their cost of money, however, is no lower than any equivalent sized and capitalized company. Leasing companies usually use higher levels of economic analysis. They know that if they get five years out of a piece of equipment, they have made a profit. Instead of trying to squeeze the last drop of value out of their vehicles (which enormously increases the management requirement), they sell or trade the equipment. In other words, they develop a financial strategy before they start the journey.

Leasing companies are **experts in equipment specification**. They tend to over-specify to reduce future maintenance costs (this reduces the need to manage the fleet as much). While this might increase the front-end purchase costs, it can dramatically reduce the life cycle costs (increasing the leasing company's profit). Many fleets cut initial costs by under-specifying and then can't understand why their breakdowns are so high.

20.2 Maintenance Costs

The lease companies see the greatest opportunity for profit in the management of the maintenance side of the fleet operation. Most leasing companies understand the need for, and advantages of, a good Preventive Maintenance (PM) system. They perform PM on equipment to avoid costs and more easily manage the fleet. You can learn from leasing companies the importance of PM and repairs.

20.3 The Full-Service Lease Alternative

- Look at the leasing company from the point of view of a long-term relationship. Do they seem stable? Can you work with their service team? Will you get adequate coverage throughout your traveling territory and wherever you operate?

- Make sure the bids are for the same equipment. The equipment bid should meet your needs. Let the leasing company bid your specification vehicle along with their own standard makes and models.

- Ascertain conditions for loaner vehicles (or lease abatements) if the unit is called in for corrective repair or PM. Include considerations for accidents.

- Look at termination clauses. Do they include terminations due to poor service, cost increases, reductions/increases in the organization's need for the fleet, cancellation. Can you purchase the units at a predetermined price?

- Determine a schedule for the predetermined, depreciated value for all units and examine the actual manufacturer's invoice.

- Determine whether you can specify your preferences such as no recaps, thermostatic fans, or radial tires.

- Who pays for cosmetic repairs and minor accidents? How is the value of the repair determined? Do you still pay for the repair if the lessor decides to forgo the repair?

- Find out if the lessor can alter monthly and per mile charges, and if so, under what conditions and in what frequency.

- Liability insurance is a major expense for any fleet. Look into who is liable and what the liability covers. Is there a "hold harmless" clause in your contract with the leasing company? A "hold harmless" clause will help protect your organization from liability suits related to accidents.

- Who pays for fuel? If you pay, does the lessor guarantee maintenance to keep the mpg to an agreed-upon level? Who is responsible for fuel taxes?

20.4 The Financial Lease Alternative

The second type of lease is a financial lease. This lease is popular in automobiles (TV commercials highlight financial leases, i.e., lease this 2010 Suburban for $XXX per month). The financial aspect of the leasing picture is one of the motivating factors for many organizations who choose a lease. A financial lease is like a loan with a balloon payment at the end of the term. Instead of paying, or refinancing the balloon payment, in a financial lease, the lessor either returns the equipment or keeps the equipment and pays the balloon (called the residual).

20.5 US Tax Law Encourages Leasing

The government determines many of the economic factors of leasing. For example, under current law, the interest on a truck loan, and the depreciation, are deductible from the organization's taxable income but the principle portion is not. However, usually all of a lease payment is deductible, making leases more attractive from a tax point of view.

COMPARISON OF A FINANCIAL LEASE TO A LOAN

You need a $75,000 Kenworth Class 8 tractor
5 years (60 payments) at 12% interest

	Loan	Financial Lease
Terms	20% down payment	First and last payment down $20,000 residual
Monthly Interest Payment	$300	$375
Monthly Principle Payment	$1000 (75,000 – 15,000 down payment)	$916 (75,000 – 20,000 residual)
Total Monthly Payment	$1300	$1291
Results		
Asset at End of Loan	Five year old truck	None
Required Cash—Day One	$15,000 (down payment) plus insurance, tax and tags	$2582 (first/last month payment) plus insurance, tax and tags
Monthly Tax Deduction	$1081 $300 (monthly interest) + $781 (eight year straight line depreciation)	$1291 (total monthly payment)

In this example, an organization that is short of cash (such as one that is growing rapidly), or one losing money (and trying to conserve cash), should choose the lease because for the same $15,000 cash investment to buy one tractor (using a loan) they could lease five tractors. An organization in a different position might opt for the loan (or pay cash).

In either case, the tax deduction would be a minor factor in favor of the lease (depending on the goals of the accounting effort).

Loss Prevention and Fleet Safety

In the long term, loss prevention is the best method to cut insurance costs and provide a safe fleet for the operators and public. Loss prevention programs do for accidents what PM programs do for maintenance. These programs identify the risks, and contributors, to accidents and manage them. Like PM programs, risk management is pro-active. These programs anticipate problems and correct them before there is a loss.

Information for this section is derived in part from the National Highway Safety Council, Smith System (a driver safety training organization) and materials provided to Pepsi Bottlers for Loss Prevention Programs.

21.1 The Safety Committee

The first step in establishing a loss prevention/safety program is to designate a safety committee. The concept and organization of the safety committee is important.

The mission of the safety committee is to:
- Determine whether an accident was preventable or non-preventable
- Determine how company policy can be modified to prevent similar accidents in the future
- Protect the operator's rights
- To involve the whole of the organization in the safety program

The US National Highway Safety Council defines a preventable accident as, "...**any** accident involving a company vehicle which resulted in property damage or personal injury regardless of who was injured, what property was damaged to what extent, or where occurred, in which the driver in question failed to exercise every reasonable precaution to prevent the accident."

Committee members might include fleet management, safety, insurance people, drivers, and management. Usually, the safety committee decides the issues on the data collected alone, without the verbal testimony of the driver (only the driver's written report). In fact, Pepsi recommends that the committee not be given the name of the driver involved to try to insure a fair hearing.

21.2 Calculating Fleet Accident Rate

The next step is to calculate your fleet's accident rate. This formula will highlight the extent of the problem and how it has varied over the years.

The National Highway Safety Council has a standardized formula to express the frequency of accidents (to establish rates and compare years).

$$\text{Accident Rate} = \frac{\text{Number of Accidents} \times 1{,}000{,}000}{\text{Total Fleet Mileage}}$$

Publish your current and historical rates with the average cost per incident.

21.3 Accident Analysis

Finally, when an accident occurs, gather all relevant information about the accident. This should include the driver's statement, witness statements, police report, medical reports on the condition of the driver (if allowed), insurance company reports, diagrams or photos, physical evidence, company investigation reports, and recent maintenance records. Gather all cost data, including repair and downtime costs. Include non-financial costs, such as the life costs to any people hurt in the accident.

Use the data collected to design a preventive program. Accidents have causes. Use the data to find out what your causes are and how to prevent them.

21.4 Keys to a Loss Prevention/Safety Program

Some people in the safety field believe that the word accident is inaccurate because it indicates that the event was completely out of human control. In fact, many accidents are preventable.

The loss prevention program must have the support of top management. For maximum effectiveness, safety should be a high priority of management.

The organization must commit resources in the form of record keeping, a safety manager (can be a part time job with other duties), and a standing safety committee.

The safety or loss prevention program should be applied across the board to all departments that operate the organization's vehicles (including sales, administration, etc.).

The focus of the safety program is people. People cause accidents. These same people are also the ones who pay the greatest cost when they are hurt. The organization should expect (and require) the cooperation of its people. The program should provide effective training, as well as written information, about any accidents that do happen (similar to the airline industry).

Ideas To Reduce Your Insurance Costs

Insurance costs are one of the larger costs of a fleet operation. There are several strategies you can use to reduce your costs while not endangering your company, drivers or the public.

- Manage your fleet to reduce accidents.
- Read your policies. This may seem surprising but most fleet managers (even those responsible for insurance) don't read their policies. Some people think the clauses are fixed, dictated by the insurance companies. In actuality, some contracts protect you better than others and some clauses can be negotiated. You may be paying for coverage which you don't need or want.
- Shop for your insurance coverage the same way you shop for everything else. This is especially true if one broker or agent has supplied coverage for a number of years. In our small fleet we saved 15% by shopping.
- Institute a driver/mechanic risk management program. Identify your risk. Take steps to manage your risk. Your insurance carrier can advise you on specific effective changes, training, and procedures you can use.
- Consider elimination of coverage that is not needed or cost effective. Look at coverage for items like towing, rental units, or other non-essentials.
- Lower the upper limit of liability insurance while insuring that your umbrella will still pick-up coverage.
- Determine how high of a deductible you can afford. Usually the higher the better.
- Consider partial, or total, self-insurance for specific coverages. This is a big area of potential savings and potential pitfalls.
- This might mean higher deductibles, or the complete elimination of collision or comprehensive coverage. Always look first at what you can afford to lose and then at the worst-case scenario.

- You can supplement self-insurance with re-insurance that sets an upper limit on your loss. Without proper thought, self-insurance can be a nightmare. Be careful of situations where you've eliminated comprehensive coverage and saved big money.
  - For example, a fire at night when all vehicles are together in your yard could be catastrophic. Or a thief could enter your yard and steal all your trucks. You're now totally responsible.
- With some planning, even that scenario wouldn't have to destroy your company. There are insurance instruments that deal with these situations. The usual solution is to purchase a stop-loss policy which starts to cover you at $200,000 or higher (but at a level you can afford).
- Be sure your drivers aren't double covered for liability insurance. They should only be covered under the truck policy not the general liability policy.
- Check that your mobile equipment (lift trucks, etc.) has the correct liability coverage. It should probably not be covered by your truck policy (for liability).
- Check to see if it's less costly to lease units that include insurance coverage.

Trading Vehicles 23

On a quarterly basis, review the reports that show comparative cost statistics by class of equipment. Units that have consistently high overall costs should be tracked through several quarters. Develop a hit list of units where action should be taken.

The hit list:
- Look at the units that cost the most per mile and/or per month.
- Ask if there is any action you can take to control the costs (alternative use, re-build, other intervention).
- Review the repair history. If you do a rebuild, are there other systems that will soon fail? If another failure looks likely, push the unit back to the retire/trade list.
- Look at the impact on availability of simply retiring the unit without replacement.

After you establish the hit list:
- These are the units to trade.
- Plan your trades well in advance. This will prevent investing a large amount of funds right before a trade.
- In some cases, it may pay better to strip the unit into parts rather than selling it whole. Be aware that these parts might only have limited life left and could cost more in labor than you would save. Consider the option.
- In most cases, it pays to keep some accessories, tires, and new batteries. Since the unit is on the hit list, whenever it is in the shop consider your options. If the hit on the unit is one month away and the unit is in the shop, look at what parts or tires you put on it. You might take the opportunity to start trading parts (as long as your activity doesn't interfere with the prime use or safety of the unit).

Vehicle and Equipment Specification

24

Vehicle specification has become a complex art where maintenance costs, fuel costs, ownership costs and downtime costs have to be calculated and balanced. There are literally hundreds of thousands of combinations possible for trucks, busses and tractors. Look at your own unique history and consider all four cost areas for each decision.

The fleet's primary job is to deliver the "goods" (including passengers, freight, school children, concrete, pallets, etc.) in the most efficient way possible. The job of the equipment specifier is to deliver the most efficient equipment for that purpose. While efficiency is hard to define over the entire life cycle, in most cases, it is better to over specify than under specify.

In the US, the million (1,000,000) mile truck is becoming standard in fleets with good specifications and excellent maintenance and operational practices. In an article in Fleet maintenance magazine, John Teig, Director of Maintenance said, "We no longer refer to our trucks as million mile trucks because that's what we expect of them." Further, he said the role of vehicle specification is essential, "all maintenance goes down to proper spec'ing." The message is that your expectation will impact your specification.

Purchase price is not the only (or even the most important) consideration. There are many managers who believe that since all trucks break down, fleet managers should buy the cheapest ones they can find, limiting their options, or worse, handing them a contract that's already been negotiated (with trucks arriving later that afternoon). All levels of government have low bid purchasing provisions, which make new equipment purchasing especially trying for the fleet manager responsible for maintenance.

24.1 Basic Questions

The first step in specification is to ask a series of basic questions about your current and proposed operation.

- What is your business? A trailer specification for moving fiberglass would be different than one moving steel, people or explosives.
- Is your product time sensitive like lettuce or lives?

- Is there talk of changes in product mix?
- What are the operating conditions?
- Are you operating in a major city, line-haul, or on construction sites?
- What are the road conditions, terrain, climate and type of traffic?
- How long is the life cycle for similar types of equipment?
- Do you trade equipment on a four-year cycle, when it is completely used up, or by a measure of utilization such as 500,000 miles?
- What are your demand hours for the equipment?
- Do you need the equipment around the clock or just one shift?
- Is the equipment used year round?
- Are there practical limitations to what your shop can handle? Consider both skills and physical limitations. The first diesels you add to a gasoline fleet will require some thought and training and you can't close the doors and repair a 48' trailer in a 45' bay (no matter how hard you try).
- How can you use the specification to increase driver productivity? A special design that saves time for drivers may be worthy of consideration. Of course, in the US, the small package delivery company UPS has made a whole business out of special designs that make the driver more productive.
- Consider your special, unique history.

24.2 Standardization

One of the most important roles of the specification process is standardizing the fleet. There are four major reasons why this is so important:

- Reduced inventory—Each type of unit requires a different inventory. In some cases, a new type of unit may even require different vendors.
- Reduced Tool Need—A new type of unit might require new tools not already owned.
- Reduced Training—When a new unit is different enough, you must invest in retraining mechanics and maybe even support personnel (rebuilders, shop helpers, etc.).
- Reduced Need for Additional Manuals, Test Gear, Programs, and Modules—You will need new sets of manuals, new modules for your "brain" readers and, in some cases, entirely new test gear.

You should have a couple of good units in your service that you change only for good reason.

24.3 Case Study – ROI and Payback

Springfield has the opportunity to re-power some older Cat engines with newer ones that are more efficient. The older engines still have good life left in them.

ROI AND PAYBACK		
Diesel – $4.00/gallon Mileage per Year – 100,000 Cost to Re-Power – $20,000		
	Old Engine	New Engine
MPG	4.5	5.5
Fuel Cost	$88,889	$72,727
Savings per Year is $16,162		
The original $20,000 investment will be paid back in 1.24 years.		
Your Return On Investment (ROI) is 81%		

At what point is this investment attractive to the company? If the older engines were due for an in-frame rebuild costing $5000 what impact would that have?

With the unit needing an in-frame overhaul the Total (net) Investment becomes $20,000 (the original amount) – $5000 (that you are going to have to spend anyway) = $15,000. At a $15,000 investment, you would have paid back the original amount in eleven months. Your ROI becomes 107%.

Consider the impact of adding $0.50 or $0.75 to the cost of a gallon of diesel fuel. When fuel prices increase you must recalculate your return on investment based on the new input.

24.4 Case Study – New Equipment Purchase

Springfield Trucking's president, Tony Windrom, decided to purchase ten Class 8 Tractors. Because of specification requirements, he has narrowed the field to Brand X and Brand Y. We will solve this by looking at past costs coupled with estimates of future costs. The estimates make a big difference since a dramatic increase in interest rates would change the outcome.

Consider the following cost areas:
- Depreciation
- Interest Charges
- Fuel Costs
- Maintenance Costs
- Downtime Costs (both hourly and per incident)

COMPARISON OF CLASS 8 TRACTORS

	Brand X	Brand Y
Cost	$74,000	$60,000
Salvage Value After Five Years	$25,000	$18,000
Interest Rate	12%	12%
Life	5 Years	5 Years
Yearly Mileage	75,000	75,000
Fuel Cost	$3.00/gallon	$3.00/gallon
Average MPG	5.1	4.7
Average Maintenance Costs	$5,500/year	$6,200/year
Downtime	30 hours/year	39 hours/year
	6 incidents/year	8 incidents/year
Downtime Costs per Incident	$350	$350
Downtime Hours	$75	$75

Which unit is a better investment?

Calculation of Life Cycle Costs

	Brand X	Brand Y
Total Depreciation for Five Years	$74,000 - $25,000 = $49,000	$60,000 - $18,000 = $42,000
Total Interest for Five Years	($74,000 x 0.12 x 5 yrs) /2 = $22,200	($60,000 x 0.12 x 5 yrs) /2 = $18,000
Total Fuel Cost	($75,000 x 5 yrs x $3.00/g) /5.1 mpg = $220,588	($75,000 x 5 yrs x $3.00/g) /4.7 mpg = $239,362
Total Maintenance Costs	$5,500 x 5 yrs = $27,500	$6,200 x 5 yrs = $31,000
Downtime Costs per Incident	6/yr x 5 yrs x $350 = $10,500	8/yr x 5 yrs x $350 = $14,000
Total Downtime Costs per Hour	30 hrs/yr x 5 yrs x $75/hr = $11,250	39 hrs/yr x 5 yrs x $75/hr = $14,625
Total Cost of Operation for Five Years	$341,038	$358,987

Brand X tractor is the better buy. The savings over five years per tractor is $17,949. In a fleet of ten tractors, that would be a savings of $179,490.

Please note (as mentioned before) that in a model, the cost of diesel and the interest rates are guesses. If the calculation is critical, recalculate the equation with different guesses and see how sensitive the model is to these variations.

24.5 Vehicle Specification/Freight Trailer Trade-Off Worksheet

Use this worksheet to determine the trade-off for each decision. Each choice has consequences in the other areas. Look at each cost area for hidden impacts. This list is just to get you started, expand it as needed for your particular situation.

VEHICLE SPECIFICATION TRADE-OFF WORKSHEET

Item	Ownership	Maintenance	Operation	Downtime
Truck class, taxes, and usage				
Diesel/Gas engine, skills, availability				
Engine governors and speed control, electronics, overall package, driver acceptance				
Clutch type/materials/brake				
Transmission type/# gears				
Propeller shafts/U-joints				
Optional dampers				
Rear axles/extended warranties				
Tag tandem/twin screw, optional				
Differential locks, liftable dead axles, bridge laws				
Air rear axles, material, comfort, freight damage				
Retarders for hilly regions, Jake brakes				
Front axles, setback axles				
Parking brakes, laws, type				
Manual/auto slack adjusters				
Mechanical suspensions				
Lighting parts, wiring system, electrical connectors, sealed systems, premium systems				
Wheels, material, construction, single cap nut systems, weight, safety				

VEHICLE SPECIFICATION TRADE-OFF WORKSHEET (cont.)

Item	Ownership	Maintenance	Operation	Downtime
Tires, radial, low profile, steel, new poly, super singles				
Tire pressure monitors, equalizers				
Hubs and oil seals material, ease, what is standard?				
Anti-spray devices				
Oregon law				
Air system dryers				
Brakes, valves, fittings, trailer timing, style, and ease of maintenance, anti-skid, friction material				
Fuel tank, size, material, weight				
Power steering, type of service air, hydraulic				
Fifth wheels, sliding mount				
Air bags, placement				
Heater, A/C, aux., electric, gas, diesel				
Aerodynamic devices				
Horns, optional location, type				
Paint, material				
Seat, mechanical, air-suspended				
Mirrors, heated				
Remote control				
Trip and speed				
Recorders, tachographs, electronics				
Air suspension, isolated cab				
Battery, maintenance free				
Amp/Hr., mounting location, mounting type				
Alternators, brushless gear/belt, capacity				
Air Cleaners, restriction, difficulty to change				
Air compressors, position of intake, freewheeling at no load				
Belts and hoses, premium				

VEHICLE SPECIFICATION TRADE-OFF WORKSHEET (cont.)

Item	Ownership	Maintenance	Operation	Downtime
Starters, electric, air				
Fuel filters, water separation, heating				
Fuel heaters, in-tank, in-line, and block heaters				
Oil filters, by-pass filtration				
Fan drives, viscous/clutch				
Fans, super light, flexible				
Engine alarm/shutdown, reduced power, time lapse				
Exhaust systems, low back pressure, single/dual				
Engine heaters, electric, fuel				
Radiator shutters				

FREIGHT TRAILER SPECIFICATION WORKSHEET

Item	Ownership	Maintenance	Operation	Downtime
Suspension air/mechanical				
Landing gear, framing shoe size, air/hydraulic location of crank				
Tire carriers				
Axles, spec to match tractor, heavy duty, XEM brake shoes				
Tandem, fixed/sliding				
Glad hand fixed/swivel				
Rear bumper, heavy duty				
Integrated floor, material, construction				
Interior liners				
Lighting system, sealed				
Threshold plates				
Door type				
Side door option				
Roof bows, non-snag type				
American Rail Road specification				
Load control systems				
Conveyors				
Reefers, diesel, electric back up				

Case Study In Alternative Use

25

Tom Duvane has eight beautiful 1998 Macks that run like-a-dream, have 450 HP engines that get 4.6 miles per gallon and have a resale value of $10,000. They are highly powered for cruising at 75 MPH fully loaded all day and night. They have been well-maintained and cost about $11,000 per year for maintenance (in a line haul usage). Company policy precludes speeding and running over weight.

In the next year, Tom will have to purchase four line-haul tractors (long distance, high mileage) and ten P&D (local pick-up and delivery-low mileage) units.

The first analysis compares leaving the old line haul trucks in service versus purchasing new. You have to make some assumptions on the life, salvage value and future maintenance costs of both units. Note that the numbers in the following analysis change every time you change an assumption (price of diesel, etc.). A dollar a gallon hike will make the replacement case more compelling, and a dollar per gallon drop, less so.

	COMPARING TOTAL COST PER YEAR	
	Old Line Haul	New Line Haul
Cost	$10,000	$78,000
Salvage Value	$6000	$25,000
Years of Life	4 years	8 years
	Ownership Cost per Year = $\dfrac{\text{cost - salvage value}}{\text{years of life}}$	
Ownership Cost / Year	$\dfrac{10,000 - 6,000}{4} = \$1,000$	$\dfrac{78,000 - 25,000}{8} = \$6,625$
MPG	4.6	5.8
	Gallons of Fuel per Year = $\dfrac{\text{annual mileage}}{\text{mpg}}$	
Gallons of Fuel / Year	$\dfrac{100,000}{4.6} = 21,740$	$\dfrac{100,000}{5.8} = 17,240$
Fuel Cost / Year	21,740 x $4/gal = $86,960	17,240 x $4/gal = $68,960
Maintenance Cost / Year	$12,800	$10,000
Total Cost / Year	1,000 + 86,960 + 12,800 = $100,760	6,625 + 68,960 + 10,000 = $85,595

There is significant profit in buying new, more efficient units.

Since your job is to recommend some ideas to Tom to cut his overall expenses based on fuel consumption, maintenance and ownership cost, you need to design a low cost plan.

Following is a case in **alternative utilization**. The old line-haul units can be de-rated to operate in the P&D fleet. Let's compare a year of projected costs for the new P&D unit to a year of costs for the old line-haul units. Again, you have to make some assumptions on the life, salvage value and future maintenance costs of both units.

ALTERNATIVE UTILIZATION

	Old Line Haul	New P&D
Cost	$10,000	$62,000
Salvage Value	$6000	$20,000
Years of Life	4 years	8 years
Ownership Cost per Year = $\dfrac{\text{cost - salvage value}}{\text{years of life}}$		
Ownership Cost per Year	$\dfrac{10,000 - 6,000}{4} = \$1,000$	$\dfrac{62,000 - 20,000}{8} = \$5,250$
MPG	4.6	5.8
Gallons of Fuel per Year = $\dfrac{\text{annual mileage}}{\text{mpg}}$		
Gallons of Fuel per Year	$\dfrac{10,000}{4.6} = 2174$	$\dfrac{10,000}{5.8} = 1724$
Fuel Cost per Year	2,174 x $4/gal = $8,696	1,724 x $4/gal = 6,896
Maintenance Cost per Year	$8,000	$6,700
Total Cost per Year	1,000 + 8,696 + 8,000 = $17,690	5,250 + 6,896 + 6,700 = $18,846

The old line-haul will save about $1,150 per year over the purchase of a new P&D unit. Using all eight Macks will save $9,200. Alternative use allows you to use these units. The debt (or cash) situation of the organization is also improved by this decision. Note that the fleet has chosen to have higher maintenance costs as a result of this decision ($10,400 higher for the next year). Of course, these costs are guesses.

Be aware that some units, because of size or length, cannot be put into alternative service.

Vehicle Maintenance Reporting Standards

The Vehicle Maintenance Reporting Standards (VMRS) project was initiated by the ATA (American Trucking Association) in 1968. The Union 76 division of Union Oil of California commissioned the initial study. After two years of study and consultation with different fleets, and discussions with other committees involved in standards, the first VMRS were published.

Since then, VMRS has gone through hundreds of revisions to keep it up to date. In fact, any carrier can apply for extensions or additional codes for special situations. These codes, once approved, will be added to subsequent additions of the VMRS.

The standard was originally developed when carriers saw a need to communicate with each other and with manufacturers. There was frustration among these groups at the lack of a consistent and easy way to exchange information. The ATA is also interested in studying the types of information that are useful to fleet managers. Their reports have been models for many in-house, and commercial, computerized fleet information systems (FIS).

The ATA VMRS standard codes allow for clear communication because they describe the vehicle and the types of information fleet managers find useful. The codes introduce a definitive description of all of the activities involved in managing a fleet.

The ATA standardized data includes:
- Equipment make, model, and specification
- Asset number
- Operator
- Type of service
- All repair data—over 8000 detailed system codes
- The parts that were used
- The failure mode—broken, bent, rusted, etc.
- Why the repair was initiated—breakdown, PM, scheduled
- Who worked on it

- How long the repair took
- When was the work done
- The cost for labor and parts
- Where the repair took place

The system also includes procedures, forms, and a report format. The goal was to have a complete system rather than the "best" system.

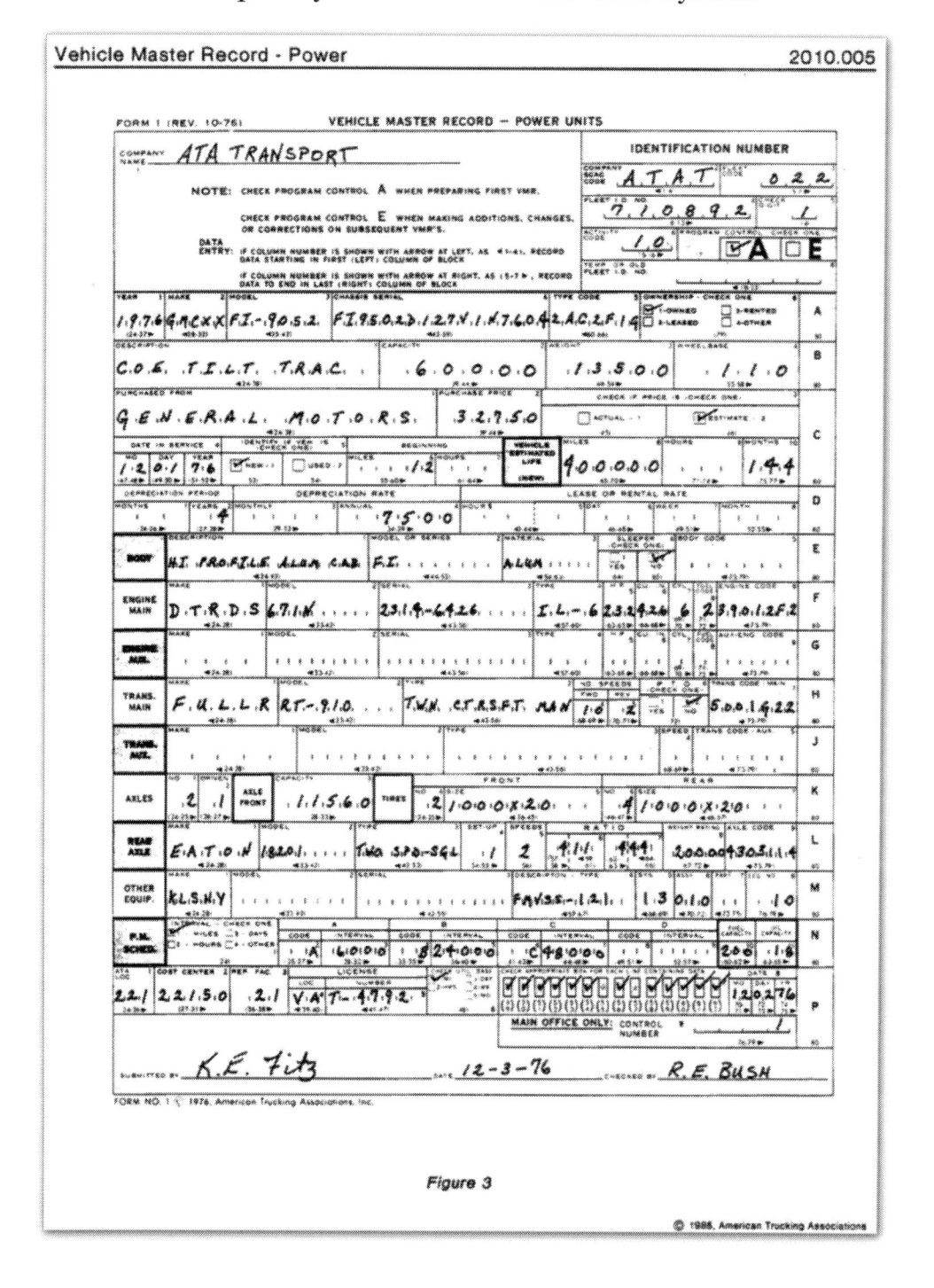

Figure 3

The data is designed in a format to expedite computerization and most computerized maintenance systems support VMRS to one degree or another. It is important to investigate any software system advertised as VMRS compatible. Many of these "compatible" systems only support some sub-set of the standard. Vendors frequently cut corners on system codes (8 digits), failure codes, and full inventory implementation. Please note that a full VMRS implementation would be a very large system and is not advisable, or practical, for most fleets. The challenge is to choose a sub-set that will serve you now and in the medium term—three to five years.

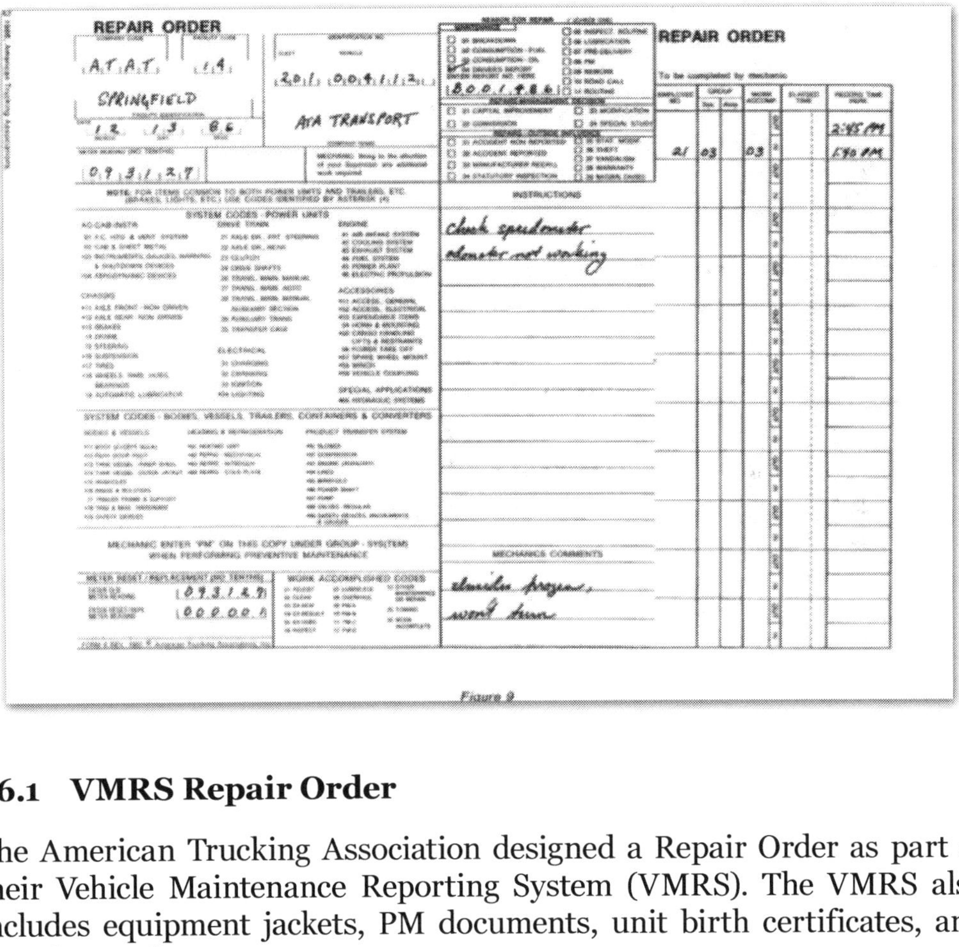

Figure 9

26.1 VMRS Repair Order

The American Trucking Association designed a Repair Order as part of their Vehicle Maintenance Reporting System (VMRS). The VMRS also includes equipment jackets, PM documents, unit birth certificates, and complete coding structures.

The author's opinion is that this is an excellent place to start. Many firms find that the full ATA VMRS has too much detail in the beginning. You have to decide how much detail you want to capture. Remember that details not captured on the RO are lost. While VMRS may not be the best solution, it is certainly a good one. It also has the advantage that many people have used it, it's the standard for the trucking industry, and most computer systems are based on it.

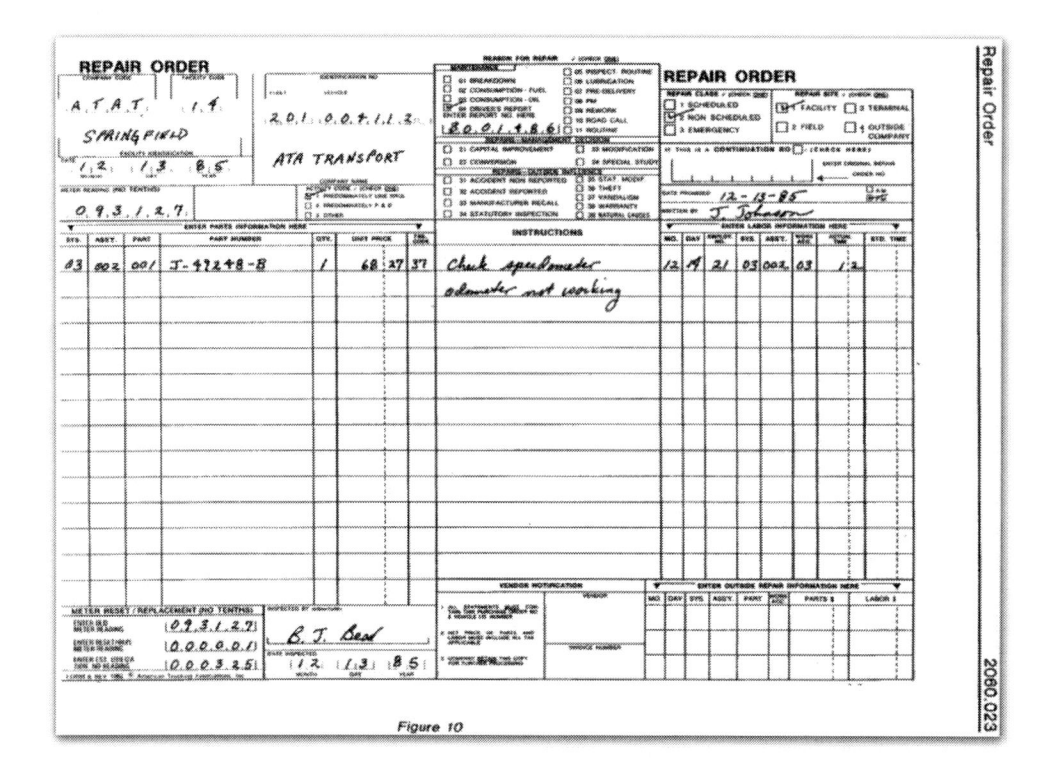

200 Basics of Fleet Maintenance

Figure 10

The Computer Generated Repair Order

27

The trend in all computer systems is to limit the amount of external paper needed to drive the system. Computerized Repair Order generation is an opportunity to cut down on that paper and use the computer to track repairs in process. There are distinct advantages and disadvantages to this strategy.

Advantages:

- Data entry is more accurate since the computer will not generate bad codes.
- Task lists that are tied to specific system or PM codes can be attached to the Repair Orders.
- Differential diagnosis items can be added to system/assembly codes. This could transfer useful information to less experienced repair mechanics.
- ROs that are lost in the shop aren't really lost.
- You can look at jobs open. This is especially true if you post hours and parts as they occur.

Disadvantages:

- A large amount of effort is spent in managing the paper flow from the computer. If a PM is required, and the equipment is not available, the RO has to be stored and "managed."
- The ease of generating ROs can lead to unnecessary paper being generated and excessive open orders on the system. In an "after the fact" system, only complete ROs or large jobs are entered.
- If what is reported turns out not to be the actual problem, it can sometimes be easier to charge the original system with the repair rather than take the time to correct it.
- Some supervisors might defer to the system generated workloads rather than solve problems as they did before (let crises occur and blame the system).

27.1 Thirteen Uses of the Repair Order (RO)

The Repair Order (RO), or Work Order, is the driving force behind any maintenance system, PM system, or analysis of maintenance function. If you are considering installation of a system and do not currently use a RO, it would be best to start one immediately and delay implementing the new system until the RO system is working smoothly.

The major problem associated with the RO is its many uses, which sometimes conflict. For example, when you consider the RO as a data entry document only, then simplicity, ease of reading, and a minimum of data to enter are its most important factors. If failure analysis is the goal, then detail is of paramount importance.

Following is a review of thirteen of the major uses of the RO and the general information each require, in alphabetical order.

1. Analysis of Failure—System/sub-assembly/part, repair reason, class of equipment, failure reason, date, and utilization
2. Authorization to Proceed—Unit number, location, estimated repair cost, and signature
3. Billing/Charge-Back Document—Labor cost, location, driver/unit number, parts cost, overhead rate, and charge/bill to department
4. Communications Document—Work to be done, unit number, task list, instructions, time/place to do work, mechanic, labor standard, and special instructions
5. Costing Document—Unit number, location, labor cost, parts cost, cost allocation code, and G/L account
6. Estimate Evaluation—Unit number, class of unit, location, system code, work accomplished, estimated labor, actual labor, mechanic number, estimated parts, and actual parts
7. History Document—Unit number, class of unit, location, labor cost, estimated labor, mechanic number, actual parts, repair reason, work accomplished, and scheduled/non-scheduled
8. Input Document—Unit number, location of repair, labor hours, estimated hours, system/sub-assembly, mechanic number, actual parts, repair reason, work accomplished, scheduled/non-scheduled, failure reason, date, and utilization
9. Labor Evaluation—Unit number, location of repair, labor hours, estimated hours, system/sub-assembly, mechanic number, and work accomplished
10. Parts Requisition—Unit number, location of repair, system/sub-assembly, mechanic number, actual parts, and work accomplished

11. Planning/Scheduling Document—Unit number, location of repair, estimated hours, system/sub-assembly, and work accomplished
12. Teaching Document—Unit number, estimated hours, system/sub-assembly, actual parts, repair reason, work accomplished, scheduled/non-scheduled, failure reason, date, utilization, task lists, diagrams, and special instructions
13. Warranty Recovery—Unit number, location of repair, labor hours, charge out rate, system/sub-assembly, mechanic number, actual parts, repair reason, work accomplished, failure reason, date, and utilization

Idea for Action—This idea is as old as the hills but worth it! Set up a file for copies of ROs involving major repairs. Review these repairs to see if there was anything that could have been done to avoid them. Write your ideas on the copy and file it. Once or twice a year, review the file and see if there are any hidden trends or ideas that you can use to improve your operation.

27.2   RO Redesign Ideas

If you have the opportunity to redesign your RO here are four ideas to consider:

- Include your staff in the development of the document. Remember they will spend more time with it than you will. Their input can smooth the way to implementation.
- Put your organization's name on top. This encourages a more professional view of the maintenance department both from within and without.
- Design your RO with wide lines. Shop pencils are usually short, stubby, and sharpened on the concrete floor. Wide lines will help with legibility when it comes time for data entry and analysis.
- Use check-off boxes wherever possible. You don't want to make your mechanic into an author!

27.3   RO Data Audits by the CMMS

All systems, both manual and computerized, rely on the accuracy of the data to make management decisions. The data has to be verified against known data points.

One of the most important functions of the data entry section is audit of the data entered. One major system goes through thirty-one tests of the data on the repair order before acceptance.

One class of tests concern the master files—those already entered for your equipment. These tests check for the following:
- Valid unit number
- Valid system/assembly/part
- Valid repair reason and location
- Valid mechanic and work accomplished
- Valid part number
- Valid fuel type for the unit
- Valid RO number

A second class of tests concern calculations or tests on existing master and detail files. These tests check for the following:
- Valid odometer reading—it should be in range based on the last reading unless it was replaced
- Replacement meter—required utilization adjustments
- Valid amount of fuel—the amount of fuel entered should not exceed the capacity of the unit
- Quantity of part—whether or not, after deleting quantity of part number, the quantity on hand is positive

After the tests are complete, certain calculations are performed using the RO data and the master files.
- Part cost from master file is multiplied by quantity on RO
- Charge out rate in master file for mechanic is multiplied by RO hours
- Total parts costs
- Total labor hours and labor dollars
- Total of all dollars including outside and misc. costs.

The tests and automatic calculations insure that accurate information is used to update the permanent detail files.

Usually the area of the program that is the most complicated is the verification of the meter reading. The meter reading drives most PM systems as well as all of the cost per mile/hour calculations.

Data entry audits—In accounting systems, it is common for the machine to generate batch totals to insure that all the data was entered and that no transpositions occurred. This batch audit technique is an excellent idea and has been carried through to certain maintenance systems. All maintenance systems should have some facility to audit the completeness of the input.

27.4 Data Entry Strategies

Many software designers have incorporated innovative strategies to facilitate data entry. Some of the ideas are very simple, such as the ability to repeat some of the previous line's data. Other ideas are more subtle, such as the ability to save the details of a Repair Order, assign them a name, and use them over and over in the future as a unit—sometimes referred to as a kit.

Every vendor of software has unique tactics and strategies to get the data from the document into the computer. They range from auto field duplication (so you don't have to re-key repetitive information), to mouse-driven point and shoot (where you highlight the selection and push the left or right mouse button).

One of the most important selection criteria for a system is an efficient data entry section, or front end. This front end is the section of the system that you will spend the most hours with. Slow, cumbersome, and unnecessary data entry strategies will cost you hours per week, every week you use the system.

27.5 Automated Data Acquisition

By far the most perplexing and continuous problem in fleet maintenance is the accurate and timely gathering and entry of data. This problem breaks down into several sub-problems.

- Accurate utilization data and state mileage—for fuel tax purposes.
- Unreadable or incomplete Fuel Log sheets and missing fuel transactions.
- Problems caused by broken meters—odometer and hour meters.
- Incomplete or inaccurate RO information and inconsistent coding of the same activity.
- Inaccurate reporting of part numbers and quantities.

There are technological solutions to some of these problems now. However, there is a continuing problem with having these various solutions talk to your maintenance system.

Utilization Data—On-board computers or Fuel Management Systems (FMS)

On-board computers can accurately monitor elapsed utilization and transmit the data to a PC. Here are some examples of the kind of data these on-board computers can store:

- Mileage
- RPM
- Speed
- Some can acquire state data—mileage by state
- Stop data
- Seat time
- Cycles of operation of equipment—for instance, compaction in refuse trucks
- Temperature of reefers

The on-board operates by reading and storing all of the inputs every interval of time—once or twice a minute. In some systems, certain inputs are tracked for thresholds only, such as high temperature, or just for time, such as seat time.

A Fuel Management System (FMS), such as the Gasboy or PetroVend units, is the more common approach to tracking utilization data. In this case, the driver enters the meter reading, and the FMS verifies the reading against previous readings before turning on the pump enabling the driver to fuel the vehicle. These fuel transactions are stored in the FMS with unit numbers, gallons, and meter readings and can easily be transferred to the maintenance system.

Fuel Log Sheets and Fuel Transactions—Fuel Management Systems (FMS)

The Fuel Management System mentioned above can handle data acquisition for in-house fueling. The FMS hardware is token operated (card, key, code) and is attached to the gas pumps on the fuel island. Every token is unique. The token has the unit number (or a number that relates to the unit number) encoded in it.

The FMS controls the electricity to the pump motor or the pump reset motor. The FMS has a transducer that is mounted in the register (the computer head—usually a mechanical device). The transducer is called a pulser because it pulses the FMS counting circuits for every 0.1 or 0.01 gallon (liter) pumped.

Meters—Maintenance software

The common problem of broken meters is usually handled within the maintenance software. Carefully analyze how a system handles broken meters. Better systems allow replacement of meters (even rebuilt meters with non-zero starting readings). The best systems will use fuel to calculate estimated utilization and add it to the unit's elapsed utilization. Be alert for systems that use the meter for elapsed utilization.

Repair Orders and Data Entry—Training and automated data acquisition

Data entry into maintenance systems consists of hundreds, or perhaps thousands, of decisions about how to handle, what to call, or how to express the work that was performed. One of the most useful methods to ensure accurate data entry is complete and ongoing training in your coding system (or an external system such as the ATA VMRS). If your mechanics and data entry people know how the system works, they are more likely to be consistent.

Automated and semi-automated data acquisition options include:

- Time Keeper Card Systems—these hand held computers (about the size of a credit card) can keep time against RO numbers, and include system codes, work accomplished, and repair reason codes. The day's work can be downloaded to a PC, which will update the open RO file.
- Laser Bar Code Scanning—all of the information on the repair order is entered into the computer by scanning bar coded pads preprinted on the RO. Parts can also be bar coded. This eliminates the need for typing and can be used very effectively in a multi-lingual environment. Products include the KeyTronic Keyboard with integral scanning wand.

Part Numbers and Quantity—Hand-held computers, laser scanning and Radio Frequency ID (RFID)

Parts costs make-up a large percentage of the maintenance dollar. Accurate data entry is extremely important to control this expense. Unfortunately, part numbers tend to be long and complicated. Inaccuracies occur when entering wrong quantities and from mistakes in physical inventory counts. Inaccurate data in any of these areas can cause stock-outs and unnecessary downtime. Opportunities for technology include:

- Hand-Held Computers—counting and logging entries during your annual physical inventory can be made easier and more accurate by taking inventory with a hand-held computer.

- Laser Scanning of Part Labels—using bar code labels and a laser wand attached to your data entry terminal can simplify and improve your parts reporting. Whenever a part is used, its label is transferred to the RO. The label is scanned along with a quantity (numbers can be preprinted on pads or the RO). Hand-held computers also come with scanning wands to aid physical inventory.
- Radio Frequency ID (RFID) tags—can be imbedded into vehicles and individual inventory items.

Computerized Maintenance Management System

All business applications have four logical components. These sections interact to become the system. The completeness and quality of the system depends on the care taken by the designers with these four areas.

When you are choosing a system, designing a system, or revamping an older system, consider these components separately.

1. Daily Transactions—includes all data entry such as RO, receipts of parts, payroll information, fuel logs, and physical inventory information. What to look for:
   ◇ completeness
   ◇ quick data entry
   ◇ logical and consistent format

2. Master Files—the fixed information about the vehicles, parts, mechanics, and organization. In better systems, the master files actually drive the report headings, utilization fields (hours and miles are never mixed) and screen headings. What to look for:
   ◇ completeness—it is very difficult to add fields to a master file after it's in use.
   ◇ storage space—not enough storage for information is a common major difficulty in a master file.

3. Processing—daily transactions are processed either in a traditional batch mode or on-line. Processing updates the PM schedule, summarizes detailed repair data for reports, and keeps all accounts current. What to look for:
   ◇ accuracy and completeness—process data through the full cycle to check that all accounts, schedules, and master files updated correctly. Most bugs occur during unusual processing conditions.

4. Demands—includes reports and screens. Large amounts of data and analysis should include reports. Inquiries should require going to print. Imagine using the system and see how it meets your requirements. What to look for:
   ◇ different ways of looking at the data

⋄ complete basic set of reports and screens

⋄ ability to alter or add reports and inquiries to suit your changing needs and growing expertise

28.1 Master File Examples

Each of these master files is an actual file in the computer. The information in these files comprises the system. Since the master file data is different for each organization, and each location for a single larger organization, each system is unique.

These files are from COSTROL, a large-scale vehicle maintenance system, reprinted compliments of ICC.

- Unit master Company master
- Computer system master Influence master
- Class master System master
- Product master Permit master
- PM master Tire master
- Reason down master Meter master
- Trouble ticket master Critical message master
- Exception master Area code master
- G/L master Cost allocation code master
- Division master Facility master
- Manufacturer master Organization master
- Type code master Inventory master
- Location master Product group master
- Product class master Vendor master
- Technician master VMRS Cross reference master
- Operator master

Master files from VEMS, a PC vehicle maintenance system, reprinted compliments of Metzco.

- Equipment master VMRS master
- Employee master Work Accomplished code master
- PM master Fixed Company Information master

28.2 Costs of a CMMS

The cost of support in stand-alone systems, that do not use your existing IT department, is a frequently unmentioned cost of in-house software. On stand-alone systems, outside IT control, the cost of support is borne by the maintenance department. On systems that run on your organization's servers, under IT control, support costs are shared by all of the organization's departments.

These costs include:
- Back-up and archives
- Communication costs, telephone lines, etc.
- Bug patches, enhancements, upgrades and any annual license fees
- UPS and power protection
- Space (office), heat, light and power
- Supplies (ribbons, toner, paper, diskettes, magnetic tape, zip disks, CD-ROMs)
- Training, all levels
- Security of data and hardware
- Labor for data entry
- Labor for audit
- Labor for analysis
- Labor for debugging and diagnosis of system/application/hardware failures and sabotage

28.3 Rule of Diminishing Returns

If you are searching for, or designing, software for maintenance, there is a rule of thumb that states you will get 80% of the benefit of a software package from the first 20% of the investment. In other words, if you expect to pay $50,000 for software, you can get 80% of the benefit from an investment of $10,000.

In the maintenance field, getting the first benefit is easy. Usually, the discipline of using any system will give you benefits. The last 20% of the benefit is very hard to capture. It requires knowledge of your operation and sophisticated knowledge of fleet analysis.

Large firms can usually justify the extra expense of going for the last percent of return because of their large numbers of vehicles. In a 1000 unit fleet, $100 per unit saved each year will fund a large system. The extra savings per unit are usually substantial.

28.4 Three Phases to Installing a CMMS

1. Planning Phase—In this phase, the project is mapped out. People are assigned. Approvals for go ahead are obtained. The computer field is narrowed to one or two system candidates.

2. Implementation Phase—In this phase, the system is purchased. Decisions based on the pre-planning information about how to handle coding are complete. Training is completed. Master files are built and audited. Data entry of any history is entered and audited. Before going live, a complete copy of the system with initial master files and data files is completed and removed from site. Live data is entered parallel with existing system. Audit completed of initial live data. Both old and new systems are run parallel for more than one reporting period. All problems are logged.

3. Support Phase—In this phase, all logged problems are addressed, either resolved or worked around. Continue the practice of doing audits of reports, inquiries, etc. Since only a sub-set of the system was installed, now is the time to decide if and when the rest of the system will be implemented. Return to planning phase for implementation of major sub-systems. Accumulate in logbook recommendations for 2.0 version.

28.5 Computerized Inventory

Normally, computerized inventory is a section in the Maintenance Management Information System. The two are very closely related. It is not very useful to maintain a maintenance inventory without also automatically costing ROs (parts charge-out cost x quantity) and reducing inventory when parts are charged.

The computer excels at applying rules and calculations, over and over, thousands of times. Inventory is a perfect example of a task that is well suited for computerization. The issue with manual inventory is discipline. Any manual system that is rigorously followed will limit stock-outs and high obsolescence costs.

When computerizing the stock room, review the elements of control:
- Require that all parts removed are recorded on ROs or equivalent document.
- The system facilitates this element by entering all items used on ROs, and charging them to units. With more complete systems, the stock room can actually charge the part to the unit and RO at the time it's given to the mechanic.

- All parts must be received, priced, physically checked, counted and signed for.
- Much of the checking, such as exact part number ordered, price, quantity ordered, etc. is very easy to do and will probably get done more often with the system.
- Parts are assigned locations. Physical inventory verifies quantity and location.
- All parts' locations are logged on the inventory system. Parts are easier to find and count for physical inventory. Systems can usually generate a physical inventory form sorted by location with the part number, description, location, and a place for the physical count.
- There must be some means for recording usage, price history, where used and substitutions.
- Usage and price history are automatically captured by the system, requiring no additional steps. Where used and substitutions can usually be entered into the part master file.
- Periodically, parts are shopped, specifications reviewed, and vendors evaluated.
- Since the system can easily generate parts catalogs, periodic shopping of higher volume items is a great deal easier.
- Periodically, part usage, fleet make-up, and availability are reviewed to adjust min/max and economical order quantity. Parts are divided into classes for different treatment.
- The real power of computerization lies in its ability to capture and analyze usage data and apply preset formulas to determine E.O.Q's. Certain systems can set-up the ABC classes of parts through analysis of yearly dollar volumes. Outside information such as vehicle retirement, and changes in fleet make-up usually have to be manually factored in.
- Parts for units out of service are reviewed for use elsewhere or disposed of.
- If the system has where-used as a data element in the part master file, it may have the ability to isolate all parts used on the retiring unit for review.

28.6 Reducing Stocking Problems

The computer system can apply the check for minimum stock level every time a part is requested. If the parts are actually ordered, then stock-outs can be reduced to the level that you set. Inventory levels can be adjusted up or down by allowing more or less stock-out conditions. Once everything is settled down, fine-tuning for seasonal variation and for age of equipment can bring inventory into line.

Many items are purchased for particular jobs and never get added to the inventory. The system can, and should, track these items and periodically report on items that start showing up regularly. You might start to stock these items to reduce costs, downtime, or both.

On the other end of the scale, you are concerned with inventory that hasn't been used. The system can easily print parts that have not been used in six months, and/or one or two years. These parts can be investigated to see if they are hard to get insurance policy stock. If they are available from outside sources, you can try to sell or trade them for usable stock.

28.7 Case Study—Computer System Planning

Tom Duvane, the Fleet Manager for Springfield Trucking, has decided to computerize his vehicle maintenance operation. Springfield Trucking has 195 power units and 314 trailers (all vans and flat beds). They are an intermediate distance, general commodity hauler. Springfield also operates a small public warehouse with six Tow motors.

Springfield operates one main maintenance facility and one satellite garage (primarily for PM). Between the two facilities there are fourteen mechanics, one gofer/helper, one stock clerk, two supervisors, and one secretary, as well as Tom.

Since Springfield had a workable RO and PM system, Tom figured computerizing would be simple. As he started to study the field, he began to have doubts. Earlier this year he started looking at packages from different vendors. He attended the ATA MSC show in Atlanta and later went to one of those two-day seminars. He now realizes that the project will take longer and cost more than he thought. He is going to have to convince his boss, Tony McDougal. Computerizing will require close planning if he expects to be on-line by winter.

Tom's questions are:
- What projects are required to computerize?
- What has to be done first?
- How much time will the entire project take?
- How much will it cost?
- Can his current staff support a computer system?

Required Projects:
1. Based on the size of the fleet, limit search to smaller systems. Send initial letters to all the system vendors published by the ATA or fleet magazines. If you want to you can get a copy of the list from five years ago and send the query only to organizations on both lists.
2. Using existing PM paperwork, create asset lists with all details. The best list will include detail to a line setting ticket level. Insure that all assets are on this list. Add shop equipment to list. Add columns to asset list for estimates of PM effort level for that asset. Enter the information into Excel, if it is not already there.
3. Inspect current ROs to see if all information on ATA ROs is collected. Add any information that is lacking. Your goal at this stage is to install discipline.
4. Concurrent with steps 1–3 open a dialog with your fleet people including the fourteen mechanics, the helper, the stock clerk, the two supervisors and the secretary. Bring in teachers from the outside to teach computer basics, if necessary. Fleet consultants can be brought in to discuss the advantages of computerization.
5. Conduct a physical inventory and clean up the stock room. Complete the A–F and Big Ticket Analysis (chapter 14). Insure all parts used are accounted for on ROs. Audit parts room paper work to see if the information is in good enough shape to enter into computer system. Look for vendors, usages, on-hand/on-order amounts, cross-reference numbers, current and last prices for all parts.
6. Conduct an analysis of all paperwork that flows through the fleet operation.
7. Review submissions from software vendors. Make up an evaluation committee with at least three people from different levels. Determine role of Data Processing department. Provide training in looking at software for committee members. Order demo or evaluation copies of all software that looks interesting.
8. Pick the best three to five vendors. Set up interviews and demonstrations. If you are a small organization you might have to travel to the vendor's site. After you locate the best package, but before you purchase it, arrange demos for your entire crew early in the process. Listen to the feedback. You might hear something you hadn't considered.
9. Purchase and install the system. Build all master files (temps might be required for this). If you put the information into Excel, import the information, then copy complete system with master files. Build a checklist of system skills such as checking unit history, entering

a RO, printing a report, checking inventory level, etc. Everyone should be given training in the appropriate level of competence with the system. Enter ROs and let everyone play with the system.

10. Wipe data clean. Enter ROs, parts orders, fuel logs, PM's, etc. Compare to your manual system. Make any adjustments. Continue for two to three periods. After this you have an operational system.

28.8 ROI and Computer Systems

A computer system can have an eventual significant impact on the five cost areas discussed earlier. Remember that the five costs areas differ in ease of impact. Operating costs are usually the easiest to lower, followed by maintenance costs, downtime, then ownership and overhead costs.

Overwhelmingly, the return from computer systems comes from the discipline imposed by the outside force represented by the computer. If you enter all your data on an Excel spreadsheet, if you recap all your numbers, if you pour over the jackets, keep reviewing the numbers and paper Repair Orders, you are running a tight operation and might not get all of the advantages of a computer system.

The most important fact to keep in mind is that computers don't maintain fleets, yet. Any impact from a system will come from action taken as a result of information provided by the system. The computer system will identify the bad actors early enough for you to have a positive impact. If you don't act on the knowledge, you don't get any return on your investment.

Don't expect to save in clerical labor by having a system, in fact, plan on an increase in clerical support, especially in the beginning. Later, you may find that the same support staff can handle a much larger fleet.

28.9 General Areas of Improvement

In addition to the specific areas mentioned on the next few pages, returns flow from improvements to the following areas:

- Improvements to labor scheduling. Only repairs where materials are available, are scheduled, information on previous jobs is available to the planner, labor requirements to complete the schedule are easily calculated and recalculated. Savings of 3% to 8% in labor.
- Improved parts availability. Less time is spent locating parts, parts can easily be kitted for jobs, and parts can be prepared ahead of time. Savings of 1% to 2% in labor.
- Increased availability of equipment. The PM system, when

followed, will reduce emergency downtime. Jobs will not start as often without adequate labor/parts to complete. Savings of 0.5% to 2% in availability.

- Reduction to inventory. Inventory is reduced through cross-referencing of parts (reducing the number and quantity of parts), automatic re-order, min/max, EOQ, exchanging old inventory for usable inventory, and ease of physical inventory. Savings of 10% to 20% in inventory level.

28.10 Early Areas of Improvement

- Within a class, sort units by MPG. Take action on the worst 15% in each class, starting with the class with the highest gallons used. You could PM the unit looking for fuel consumption items, re-train the driver, move the unit to lower mileage service, replace tires with radials, etc.
- On the same review, note vehicles with excessive fuel use and which drivers are involved. Shift the drivers and follow their fuel use. If a pattern develops, you may be dealing with theft of fuel. On some sites, theft accounts for 5% to 10% of fuel used.
- On vehicles that are still under warranty, collect all warranty claims and analyze the component systems that were hit. Where one component system has had a high frequency of failures, apply for warranty adjustment for all of the vehicles that were purchased together for that system. You may be able to get either additional warranty recovery, or some other special consideration (extended terms). Note that component manufacturers and truck manufacturers reserve a percentage of the value of the sale for warranty (typically 1.5% to 2%).
- If you added your history when you started the system, and you feel that the cost information is accurate, generate a report that shows total cost per mile within each class. Target the worst 15% for action, such as retiring the unit, slating the unit for rebuild, or moving the vehicle to less strenuous service.
- Review inventory parts as you set up the master files. Look for parts for vehicles that are no longer in service. Note the parts, or put them aside for sale or trade. This not only generates funds for investment in needed parts but also frees up space.
- Review all vehicles during the first few months of the system looking for units without mileage/fuel (no activity). Some systems have No Activity reports. Many organizations have a couple of units parked against the fence for emergencies. Unfortunately, usually

when they are needed they don't quite run. Retire, trade, or scrap those units and lease emergency units when you need them.

- Review the history of any unit requiring a major repair. Before you start the repair, decide where on the critical wear curve you are. You may want to consider trading the unit depending on the frequency of repairs, emergency breakdowns in spite of PM activity, degeneration in the CPM over a period of time, and other systems showing signs of failure.

28.11 Long Term Areas of Improvement

After the system has been installed for at least a year, preferably two, other returns are available. Analysis of your mass of data will uncover the specific facts of your situation and environment. All areas of cost can be looked at, analyzed, manipulated and compared.

- Adjust PM task lists by failure experience. This will maximize your PM effect per hour.
- Review failures to improve new vehicle specification.
- Create experiments with additives, tires, farings, synthetic lubes, bypass filtration, foreign makes, etc. Numerically know the result of your experiments.
- Tracking productivity of mechanics will help pinpoint those who need training, rewards, discipline, or those who need to be removed from the workforce.
- Look for inventory that has not been used in one or two years to see if it belongs with vehicles that were traded.
- Isolate costs of satellite operations for cost effectiveness. See if they might cut the overall cost of operation.

28.12 Case Study—Effects on the Bottom Line

Tony McDougal is the president of Springfield Trucking. He's been in the trucking business since 1989 and has seen a lot of changes with deregulation, OPEC, and conflicts in the Middle East. He is a little uneasy about the continued cost of fuel and his ability to raise prices fast enough to cover costs. Tony prides himself on the fact that Springfield is a profitable trucking company with the net about 7% of sales.

Nancy Strathmore, Springfield's sales manager, has proposed an expansion into some specific commodities. This expansion would require $94,500 in new equipment. The expansion would bring in $450,000 in new sales revenue per year.

Tom Duvane, fleet manager, has proposed investment in a new computerized PM system. Calculations show the return will come from reduced fuel costs and reductions in maintenance costs. Tom believes the system would allow his existing staff to support more equipment. The investment is $75,000, with returns of $25,000 in year one, $50,000 in year two and subsequently, at present utilization figures.

Tony is not inclined to make both investments in the same year. He needs to know which one should be done first.

	COMPARING THE BOTTOM LINE	
	Sales Department	**Maintenance Department**
Investment	$94,500	$75,000
Extra Revenue	$450,000	
	Return per Year = extra revenue x net profit percentage	
Return per Year	450,000 x .07 = $31,500	(25,000 + 50,000) /2 = $37,500
	Years to Pay Off Investment = total investment / return	
Years to Pay Off Investment	94,500 / 31,500 = 3	75,000 / 37,500 = 2

Sales Department Investment:

Return per year (extra revenue x net profit percentage):

$$\$450,000 \times 0.07 = \$31,500$$

Years to pay off investment (total investment/return): 3 years

Maintenance Department Investment:

Years to pay off investment: 2 years

Which is better? Maintenance Department Investment

One big question is a financial one. What would be the impact of leasing the special equipment? Could the payment be made with the $31,500 return?

How is Tom Duvane going to compete with the sales department, MIS, and other departments? Should he invite Tony to the shop to review the numbers, hire a PR firm to sell his point of view or invest in graphics software to make slick presentations?

Driver Vehicle Inspection Reports

The Driver Vehicle Inspection Report (DVIR) is required in the US for commercial vehicles over 10,000 pounds. One DVIR is prepared at the start of a shift and another one at the end of the shift. The DVIRs are given to the dispatcher, who insures there is 100% compliance, notes problems, and passes the reports on to maintenance.

Some of the areas that might be covered by the DVIR include:
- Brakes/air compressors—all types
- Steering—including symptoms caused by wheel alignment/balance
- Lighting/reflectors/mirrors, tires/wheels/rims/valves
- Horns/ hazard flashers
- Coupling devices—king pins, glad hands
- Fuel tanks
- Heater/AC
- Ladders/ access systems
- Engine problems
- Emergency equipment

This is primarily a safety communication tool to insure good vehicles are given to drivers, with prior problems repaired, and to insure that when the vehicles experience problems, it is communicated to the maintenance department. The driver signs off on the DVIR to show acceptance.

The DVIR is a challenge to maintenance. If a driver writes one up there had better be a work order, or other inspection ticket, looking at the area mentioned by the driver. If there isn't, the fleet is in big trouble with the US DOT (Department of Transportation).

Each state in the US has slightly different rules for the pre-trip inspection required under their CDL.

For the MD pre-trip inspection training booklet with details of how and what to inspect, please visit:

> *http://www.mva.maryland.gov/Resources/DL-152.pdf*

For the CA DOT pre-trip form, please visit:

> *http://www.dmv.ca.gov/forms/dl/dl65prt1.pdf*

Using Statistics To Identify Problems

What is the Normal Curve and what aspects of it are important to us?

There are hundreds of curves or distributions to represent events in the world. The most useful one is the Normal Distribution. The Normal Distribution curve can help us identify the bad actors in our fleet, among our mechanics and in parts used.

The Normal (or Bell Shaped) Curve is a graphic representation of a large number of similar readings. The readings tend to be more frequent the closer you get to the sample mean (called X in mathematics). This sample mean is also the average reading. One of the several useful properties of the normal curve is that it is symmetrical on both sides of the X (average point).

Many of the advantages of the normal curve can be shown graphically without mathematics. We will review both methods. The mathematical approach could be used to prove a case that might involve large sums of money.

You can divide a normal distribution into partitions that are extremely useful to help you manage your fleet. The size of the partition is called one Standard Deviation (SD). The useful property of the SD is that 68.27% of your readings (or data points) will be within 1 SD of X (that is X±SD) and 95.45% of your readings will be within 2 SD of X (X±2SD).

CALCULATING STANDARD DEVIATION
$X = \dfrac{\text{the sum of all your data points}}{\text{total number of data points}}$
Difference = each data point - X
Difference^2 = Square the difference of each data point
$\text{Variance} = \dfrac{\sum D^2}{\text{total number of data points}}$
Sum the difference squared for each data point and divide by the total number of data points
SD = $\sqrt{\text{(Variance)}}$ = the square root of the variance

The readings can be cost per mile, miles per gallon, flat rate hours completed per week, or any numerical measure. In the real world, when you plot readings from any measure, your data will approximate the normal curve.

How can the mathematics of the normal curve help manage your fleet?

The Normal Distribution Curve is the mathematics behind the important and commonly known management rule known as the 80/20 rule. This rule states that 80% of your problems come from 20% of your units, people, parts, etc. We call these 20% the bad actors. Our job as maintenance managers is to identify and fix these bad actors.

30.1 Using the Normal Distribution

If the class (like vehicles in like service) is well chosen, there are no other factors and the partitions reasonably selected, the curve will look like the normal distribution. Events in the real world will tend to look more and more like the normal distribution as the samples get larger. In this real life example of thirteen buses, the curve has another hump on the low MPG. There may be another factor contributing to the curve.

Remembering that the normal distribution is symmetrical around X and that 68.27% is within 1+SD, you can identify the units that have to be managed. 15.9% of your fleet is more than 1 SD lower than X. These 15.9% are your bad actors. These are the units that are causing your problems and costing you money. The best return on investment will come from devoting attention in this area.

The MPG readings are:

5.62	5.11	4.86	3.81	3.05
5.57	5.11	4.55	3.58	
5.13	4.93	4.36	3.10	

The X (sample mean) = 4.52 MPG

The SD = 0.7246

The lower boundary = X − SD = 4.52 − 0.7246 = 3.8

Vehicles with fuel consumption below 3.8 MPG are of interest. In this case, that means the units with the following MPG—3.05, 3.10 and 3.58.

Improvements to the performance of these three units will have maximal impact on your fleet operation costs. Using averages calculate the gallons saved by improving these three units to the mean.

Miles traveled for these three units = 21,584

Average MPG for these three units = 3.24

Gallons of fuel used by these three units = 6691

Gallons of fuel used at 4.52 MPG = 4775

Potential saved gallons = 1915

No other group of vehicles will give you as good a return on your time and money investment then those more than 1 SD below the X.

Note—The new average MPG would increase to 4.81, an increase of 6.1%, just from managing these three units.

30.2 Statistical Failure Analysis

Failure analysis is a statistical view of the incidents in a fleet's operation. It reviews the failures and comes to some conclusions about their frequency and cause.

The technique of failure analysis is to determine the elapsed utilization between incidents of failure. You are trying to gather failure information in a form that lends itself to ready analysis.

Use Failure Analysis:
- Failure analysis is the major tool in the establishment and updating of PM systems. If failures are too high, then increases to the task list or in the inspection frequency are required. If failures dip too low, then the reverse is true and too much money is being spent on the PM activity for that component.
- To set up a PCR program
- Use the information to compare two makes of components. You might want to compare Eaton to Bendix components to choose one over the other.
- Failure analysis can prove experiments and provide data for efficient decision making. An example is looking at engine failures for synthetic versus natural oils.
- Fleets constantly evaluate their specification. Failure analysis can help improve specification.

BELT FAILURES ON THERMO-KING REEFER UNITS		
FID#	FMTK648	
Incident Number	Hour Meter at Failure	Elapsed Hours
1	234	234 - 0 = 234
2	631	631 - 234 = 397
3	1301	1,301 - 631 = 670
4	1982	1,982 - 1,301 = 681
Unit Average = 495.50		
FID#	FMTK700	
Incident Number	Hour Meter at Failure	Elapsed Hours
1	991	991 - 0 = 991
2	5152	5,152 - 991 = 4,160
3	5424	5,424 - 5,152 = 272
4	5897	5,897 - 5,424 = 473
Unit Average = 1474		
Unit Average without 4160 reading = 579		

30.3 Normal Distribution and Graphic Representation of Failure

The normal distribution is a graphic representation of a function or statistical universe. These graphic techniques are powerful tools for the analysis of an unknown distribution.

In theory, failures of components should follow the rules of the normal distribution for each failure mode. Once you decide to analyze a particular failure for a particular class of equipment you have to gather the data (the form in the example is elapsed utilization between incidents).

The second step to graphing the chosen failure events is dividing the elapsed utilization between incidents into reasonable partitions. The numbers of failures you collected and how closely clumped the numbers are will indicate a partition size. These partitions should be wide enough so that the partitions around the mean have many elements. The partitions shouldn't be so wide, however, to exclude reasonable analysis.

The third step is to start adding the failures to the correct partitions. The curve that results will look something like the bell shaped curve. You can estimate the mean, standard deviation, and mode visually, use the formula, or use a statistical calculator. These numbers will help you setup a PCR program.

```
         Graphic Representation of Belt Failure Incidents

 3                                                              X
 2                             X                                X
 1         X         X         X              X                 X
   ─────────────────────────────────────────────────────────────
   0   250   500   750   1000   1250   1500   1750   up
   Elapsed Utilization
```

Analysis would show:

Average Hours = 1,021 (with the last incident)

Average Hours = 575 (without the last incident)*

Standard Deviation = 784 (with last incident)

Standard deviation = 175 (without last incident)*

Mode is above 570, less than 670 (equal number higher and lower)

*Many statistical algorithms (methods to solve problems) smooth the function by discarding the highest and lowest readings.

In this case, the elapsed hours are the hours of operation between repair incidents. If you had hundreds of incidents of belt failures on TK units, the average and standard deviation would start to be statistically significant.

Failure analysis is a useful tool when your fleet has had hundreds, or thousands, of failures of the component being considered. There are other limitations to failure analysis. The components being analyzed must be close to the same or identical.

In this example, adding in data from alternator belt failures may skew the analysis and make the outcome meaningless (although a study of all belts might have some use, especially if you are considering the effect of the newer belt materials).

The units should be similar and the severity of service should be similar. For this reason, limit failure analysis to class (like units in like service). The goal in any statistical analysis is to isolate and factor in any outside influences that might affect the analysis.

The goal here is to investigate the unknown statistical universe of elapsed time for failures of belts on TK units. We ask the question, "What is the probability of failure of TK belts after NNN hours?"

Divide classes by lifetime utilization for greater accuracy. Division of the units into lifetime utilization is another point that affects the outcome and might lead to higher levels of accuracy and usefulness. In this case, we break-up the incidents by the total lifetime utilization on the unit at the time of the incident. There is an indication that failures of new components such as belts (or alternators, starters, engines, almost anything) have different failure curves based on the age of the unit they are going into.

There are many ways to measure a fleet's performance with regard to its cost effectiveness and efficiency. We also introduce methods to capture data to measure and quantify the status of the fleet.

Fleet Key Performance Indicators

We will introduce five important concepts:
- Cost category identification
- Measurement of fleet operational improvements
- A typical standard for maintenance reporting
- Maintenance record keeping

31.1 Measurement of Fleet Performance

Metrics—measuring effectiveness

Before any changes take place, before you install any new computer system, before you improve your new vehicle specification, you must measure your flect's current effectiveness and performance. These financial and other numeric measures could be trended on a monthly basis and summarized yearly.

Proving to top management that a month's performance is actually an improvement, even when some of the costs go up, is one use of trending. In previous sections, categories of costs were identified. You saw how the five cost areas are interrelated. Improvements in fleet performance frequently show up as increased expenses in one area (the investment) and reductions in the future costs in another area (the return).

You can also use trending to learn if a particular set of performance numbers are an improvement, or part of the normal variability of that measure. Watching the trend of the number will quickly tell you if you are improving, or just moving back and forth.

31.2 Ratios and Numbers to Measure Fleet Performance
- Cost per mile (CPM)
- Cost per hour (CPH)
- Cost per utilization unit

The primary measure of a fleet's performance is cost per mile. This measure can be broken down to the five cost areas (even into some of the sub-areas). CPM is a useful tool when broken into divisions (comparing different groups within your organization to each other, assuming that their equipment is similar). As a fleet management tool CPM is useful when the equipment is divided into classes of like equipment in like service. The worst units of each class can then be managed.

In construction, material handling, and stationary equipment, cost per hour is the primary measure of performance. The comparisons and uses are the same. The general expression is costs per unit of utilization.

31.3 Cost per Secondary Utilization

The primary utilization of most vehicles is miles or hours. The secondary utilization is how the unit is used in terms of production. This measure is closely related to CPM and is used widely by commercial fleets. For example, in the beverage industry, costs per case (soft drinks, beer) are a valuable measure of a fleet or a particular route. Any output such as ton, ton-mile, people moved, cubic yard, even revenue dollar can be used as a measure of fleet performance.

31.4 Maintenance Costs per Asset Value of Fleet

As your fleet ages, this number goes up rather rapidly. Both the asset value is declining and the maintenance costs are increasing. If you have a policy of rotating your fleet regularly, so that the average age stays the same, then this measure can be useful. This is a leading indicator (changes will occur early) of fleet condition and age.

31.5 Maintenance Cost Index

This index measures total current month's actual cost to the same month in the previous year or two. The current month should be in-line with the previous year, within the variations in fleet size/make-up and inflation. Significant differences may be cause for investigation.

31.6 Actual Cost Compared to Budget

This provides a check on your budget. Variances could reflect a problem with your operation or with your budget.

31.7 Ratio of Indirect (Overhead) to Direct Costs

There is a conceptual optimum ratio for each organization. The idea is that if the overhead to direct cost ratio falls too low, then not enough planning is taking place and overall costs will be high. On the other hand, too high a ratio indicates wasted money since only a certain amount of overhead is required to manage a fleet of a given size/make-up.

31.8 Direct Labor Effectiveness

There are two related measures of labor effectiveness. The first is simply payroll hours to repair order hours. How much of the shop time is spent on assigned work on the fleet. The second measure looks at payroll hours compared to a fair days work. In this measure, the labor standards (or flat rates) for all jobs completed are added together (regardless of how long they actually took) and compared to the amount of hours paid for.

31.9 PM System Audit

There are two measures that help verify that the PM system is in place. The first compares Inspection/PM hours to all direct hours. The result of an effective PM system is that repairs are uncovered by the inspector and scheduled. The second measure compares total PM hours to total direct hours.

31.10 Emergency Hours

This is a total of the number of hours spent working on jobs that interrupted your daily schedule. It is expressed as the percentage of emergency hours to total direct hours. To a certain point, lower percentages indicate an effective PM system. If this falls too low, your organization may be over-investing in PM.

31.11 Unscheduled/Scheduled Overtime

Tracking of both types of overtime can be a useful judge of your crew size and PM system. No scheduled overtime usually indicates an over-crewed shop. Small amounts of overtime (up to 2–3%) are normal. Excessive non-scheduled overtime usually indicates an ineffective PM system.

31.12 Downtime Hours/ Downtime Reasons

Downtime hours are hours lost due to equipment being out of service for any reason. The ratio is the percentage of downtime to total time available (units x time per day x days in reporting period). Maintenance related downtime must be separated from other types of downtime: Accidents, Abuse, Legal, recall, Re-build, Sabotage, Schedule/Route changes. The downtime ratio will improve as your ability to predict failures, before they occur and correct them, improves.

31.13 Repairs Delayed/Delay Reason

This represents the total amount of approved and scheduled work delayed. This number represents work that cannot be completed. Reasons include Awaiting Material, Special Tools Required, No Labor Available, or No Place To Do Work. Excessive delays indicate problems with planning, stockroom, or crewing. Whenever a job is delayed, for any reason, time is lost.

31.14 Rebuilds and Capital Projects

While capital projects are frequently given to maintenance, they are not the primary mission of the maintenance group. This number (in dollars or hours) could indicate that maintenance is being deferred to work on capital projects. Many shops mount their own truck bodies, for example, and charge their time to the maintenance budget. Top management then wonders where the funds went.

31.15 Road Calls

This is a direct measure of the effectiveness of the PM system. These are usually the most expensive and disruptive types of repairs. Distinctions should be made between Mechanical, Out-of-Fuel, Accident and Tire Road calls. As the PM system matures you will see a reduction in the quantity and severity of these calls.

31.16 Repeat Repairs

This is a measurement of the effectiveness of your repair labor. It is expressed as a number of incidents and total hours. Standard definitions of repeat repairs should be determined.

31.17 Technologies

On-Board Computers

General information—On-board computers can now monitor all of the activities of the driver, tractor and load. Driving patterns and abuse account for 10–20% of the cost of operation of over the road heavy equipment.

There is an even greater potential savings in the area of driver productivity. Drivers are one of the last great un-tapped areas of productivity improvement. Accurate data about actual driving and unloading times are the basis for control of the drivers. Exciting technology is now available to address these issues.

These on-board computers have different specifications. The early units could only record RPM and MPH. Later models have input capabilities for:

- Vehicle Data—RPM, MPH, engine hours, elapsed utilization, actual fuel consumption, cycles (of chassis mounted equipment such as a trash compactor), PTO hours and oil/water/load temperatures
- Driver Data—driver in-cab input, seat time, idle time
- Trip Data—stop identification, stop sequence, stop arrival/elapsed/departure times, amount unloaded, invoice generation, route settlement, crossing of state lines, state mileages

Some systems have capabilities for alarms/shutdown to reduced power levels and cruise control. Other inputs are being added for specialized use. Specialized devices are available for just alarm/shutdown or fuel consumption.

How On-Board Computers Work

- Input—The vendor has designed sensors (officially called transducers) that give off output that the on-board computer can understand. These sensors can be as simple as relays whose coils are across the ignition system to determine hours-on, to transducers that generate variable frequency proportional to the road speed. The trend is to use standard sensors already in use for the standard data collection.
- Execution Loop—Each recorder has a timing loop which determines how often each of the inputs is read. Each time the loop cycles, the inputs are recorded. Loop cycle times vary from 30 to 120 seconds.
- Interruption Handling—Most on-boards can handle interruptions from special events such as engine alarm situations (low oil pressure). Other interruptions might include driver input.
- Storage—Each reading is stored in the on-board computer. In some cases the data is compressed as well as stored.
- Transfer—When the unit returns to the terminal (or shows up at any terminal) the data is unloaded. Unloading is physically accomplished through a data module, wireless LAN connection, or umbilical cord.
- Post Processing—All of the systems supply software to turn the data into usable reports and inquiries. The actual usefulness of the reports varies greatly from vendor to vendor and also depends on your needs.

- Installation—The computers are mounted in a steel box about 12" x 6" x 4" deep. The box is connected to the un-switched side of the battery. Most of the time is consumed in fitting the senders and routing the wiring. Usual retrofits take about eight hours. Most units can be ordered already mounted on new trucks.
- Sample Reports Available—Report on all aspects of a day's drive including idle time, over/under rev times, over speed, stop data, etc.
- Driver Input report includes state miles, detailed stop data (like delay codes, quantity, etc.), fuel purchases
- State/province miles – KM/gallons – 100L report
- Time at stop report relates time at stop to units unloaded and delay codes
- Speed violation report, RPM violations, Idle time violations
- Minute by minute analysis of speed, RPM before accidents

Routing Systems   32

Routing systems try to minimize mileage, time and/or delivery vehicles. All of these improvements have an impact on costs of fleet operation. Many costs (notably operating costs) are directly proportional to utilization. The impact from changes to the routes is felt immediately (the next time the unit fills up with fuel and goes an extra day).

When installing routing systems, you can frequently reduce the fleet size by several units. If you use the technique of retiring the bad actors discussed earlier, you can lower both your ownership costs (fewer units) but also your average operating costs (by getting rid of the high cost units).

There are several sub-problems to the overall routing problem. The sub-problem is usually dictated by the type of delivery of services. Most fleets have components of all of the types.

1. Many fleets, such as school buses, mass transit, grocery, newspaper, beverage, dry cleaning, and many others make regular stops. They have well defined customers who get serviced on a route. The route generally evolves over time or is re-set yearly.

2. These fleets have a well defined customer base but the deliveries are based on orders entered (rather than, for example, a regular 3 PM stop). Organizations in this category include some of the above and retail oil delivery, Para-transit, wholesale, industrial/commercial service organizations, and others.

3. Firms whose day to day schedule varies and whose customer base is very large such as package delivery (package pick-up is usually more like the first group), common carriers, retail service, most salespeople, taxicabs, etc.

Within all of the types, the lower type problems are sub-sets of the higher types. The solution to the type two problem would work for a type one problem. At the core of the type two system is a kernel with a type one system in it.

Type 1 fleets usually need a package that will sequence the stops to minimize the mileage/time. Usually routes are chosen based on maximum load, delivery windows, or elapsed time. In one case, an inexpensive package was used to re-sequence the stops. The stops per route are coded into the system using latitude and longitude or distance between stops (you manually measure actual distances). Hitting the stops in the suggested order saved 9.3% of the day's mileage. Issues like loads, equipment type, and windows are usually externally handled. This is a utility rather than a live on-line routing system. These types of systems usually cost $350 to $1000.

Type 2 fleets need some level of batch load control and routing. Based on input from order entry, you must first divide the day's business into loads which are constrained by cube, weight, elapsed delivery time, delivery windows or special equipment requirements. After the day's loads are selected then that stop sequence is dropped into a type 1 kernel to optimize the stop sequence. The complete customer base has to be coded into a large data base with the exact locations and special requirements (delivery windows, special equipment needs).

Type 3 fleets require coding of the streets and locations in a huge data base. These data bases are available from agencies and from the Type 3 software vendors. In addition to the requirements above in Type 1 and Type 2, a Type 3 system must be able to locate customers based on address and/or zip code. After the customer is located and their special needs are determined, they are loaded into a Type 2 program. The largest of these systems have actual streets, addresses, and intersections coded into the data base. Many use custom forms off the Internet based Mapquest.com or Google maps.

Vehicle Locating Systems

Systems are now available to locate vehicles using GPS navigation satellites and cell technology to transmit coordinates to the dispatch center. The dispatcher can locate the vehicle on the map in real time. The vehicle stays in the center of the display and the streets re-orient themselves as the vehicle moves and changes direction.

The locating system uses navigational satellites to locate the unit. The GPS keeps querying the satellites and correcting the position. Streets and points of interest are coded onto maps that are displayed on the GPS unit mounted in the cab (and transmitted by the cell network back to the office). It is also used to keep drivers from getting lost. This technology has excellent applications for taxi, police, fire, salespeople, and local delivery type fleets.

One fleet locating system is linked to the on-board computer so that you can not only locate the vehicle but immediately see what is wrong.

GM's On-star system for passenger cars combines this with two-way cell phones. The operator can talk to the vehicle's occupants and discuss the problem (the attendant can see if an air bag has been deployed and send police or medical personnel). They can also send unlock codes in case you are locked out of your car.

Vehicle Identification Systems

One of the continuing problems is the identification of vehicles to computerized fuel and data acquisition systems.

The most popular methods are:
- Cards—Magnetic stripes, optical, weingard, smartcard, bar code
- RFID (Radio Frequency ID) chips—These are permanently mounted somewhere in the vehicle
- Keys—Data key, keycard, mechanical
- Data Modules—Rockwell, Engler, etc.
- Key pads—Just type in the unit number

Wireless 35

Wireless communications with the truck's brain is revolutionizing data collection. A truck can come into the yard and be scanned for trouble codes, mileage, RPM history, and other operational facts. Wireless communications are related to on-board computers but go much deeper into the trucks electronics since it talks directly to the OEM brain.

Currently wireless interfaces can talk to the diagnostic system (engine, transmission), ABS braking system and some accessories.

Fifty Notes To Take With You

1. Review the questionnaires. See where you are today and what good maintenance practice is.
2. Be sure your staff knows what you expect of them in their job.
3. All impacts to costs of fleets are impacts to the five cost areas—ownership, operating, maintenance, overhead, and downtime.
4. The five cost areas interact. Watch out for interventions that sacrifice one cost area for another. Be sure there is an overall improvement.
5. The sum of all costs over time is the life cycle cost. This is where results should be viewed.
6. To know how your fleet is doing, you must manage using critical ratios and numbers. Feeling that you're fleet is doing ok is not enough anymore.
7. A good source for information, standards and contacts is the ATA (in North America).
8. Agreement on vocabulary is a first step in getting everyone to understand each other. It also helps new hires.
9. Know your resources and the demands on your resources.
10. Know your formal and informal work rules.
11. Know your fleet.
12. Know your inventory.
13. Use of Gannt, CPM, and PERT charts will improve your ability to manage projects.
14. Review paper flow to minimize unnecessary and duplicate paperwork.
15. Keep alert for tools for fleet managers from outside the industry, including advances in systems, project scheduling techniques, investment analysis and statistical analysis.
16. Computer systems are not a universal panacea. They can provide the outside discipline to do the work necessary to control your fleet.
17. Computer systems can do the drudge work of maintaining your

fleet system to a useful level, given the input. This assumes you read and act on the output.

18. Pay attention to the form and layout of your repair order because almost all analysis starts there.
19. The RO is used for over thirteen types of activities, analysis and authorizations.
20. A labor hour that seems to cost $20.00 actually costs over $45.00.
21. Only 30 – 50% of your mechanics are at work at any given time, unless you manage the lost time.
22. Use of standards will increase your productivity and the smoothness with which work gets done.
23. Use Chilton, Manufacturers, and third party publisher's flat rates to get started with scheduling. Temper these flat rates with your own observations (RE).
24. Prepare a yearly labor budget to find out where you are in relationship to what you need.
25. To control your shop, you must control the repairs, hour by hour, mechanic by mechanic, and part by part.
26. The key person to a successful schedule is the planner. Pick them with care.
27. Review for the symptoms of an inventory out of control.
28. A $10 part actually costs $12.
29. Review big ticket items to cut total inventory.
30. Review "A" level items to cut money spent on materials and increase your productive use of money.
31. Proper parts purchasing includes buying at the best tier. It can cut your costs.
32. Parts can have several numbers. It pays you dividends to know the interchanges.
33. Manage the critical wear point and you will manage your failures.
34. PM is a system with feedback. Feedback can reduce costs by putting the inspection dollars where the failures are taking place.
35. Choose items for your PM list carefully because any item on the list can represent a hundred or more hours per year (which could be wasted if the PM is not well chosen).
36. Use PM to help fight the grease-monkey concept of fleet maintenance personnel.

37. Early detection avoids core damage.
38. Under-funded past sins will haunt any PM effort.
39. Failure analysis can help set PM intervals and tasks.
40. For reduction of emergency repairs, use PCR.
41. Methods are available to predict failures so you can intervene at the most economical time, balancing extended life against core damage.
42. There is a mass of data available to cut fuel consumption.
43. Fuel tax is often an invisible additional cost of fuel. It can also be managed.
44. You can save money, time, and fuel by using in-house fuel with computerized FMS.
45. Your first shot at reducing life cycle costs is proper specification of the vehicle you purchase. Work and re-work your financial models to maximize return.
46. PM can increase the availability of units, and reduce the number of units you need.
47. Plan your trades well in advance.
48. Vehicle abuse and driver productivity can both be monitored by on-board technology.
49. Route improvement will pay immediate dividends in reduced operating costs and future dividends in smaller fleet size and reduced maintenance costs.
50. New technology such as vehicle location devices are making managing fleets a new ball game.

Thanks and Good Luck!

Joel Levitt

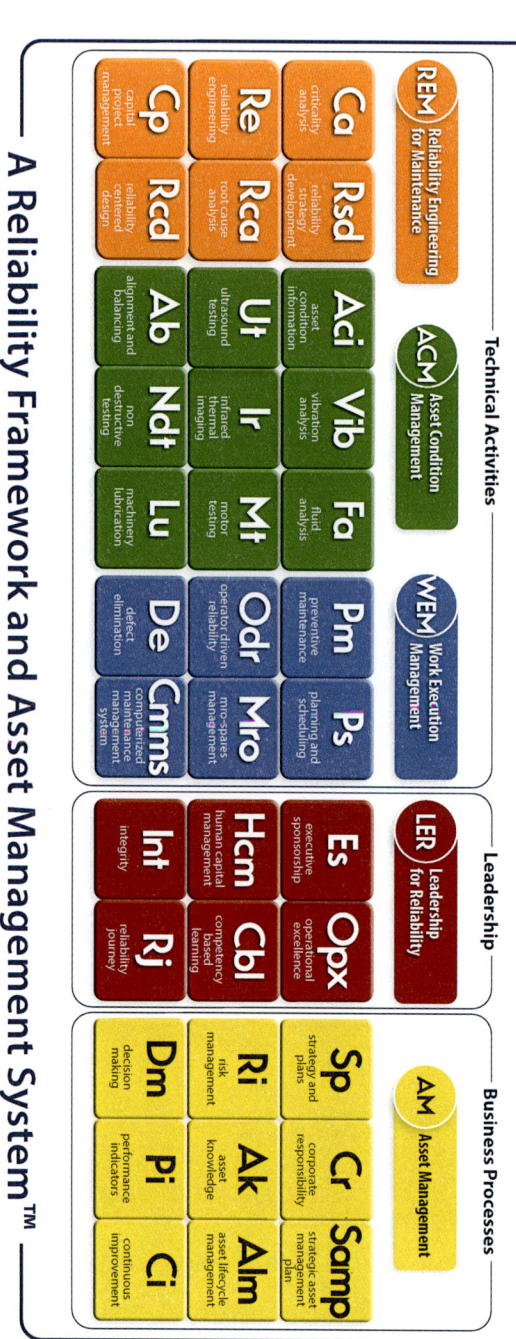

Uptime® Elements®

A Reliability Framework and Asset Management System™

Technical Activities

REM — Reliability Engineering for Maintenance
- Ca — criticality analysis
- Rsd — reliability strategy development
- Re — reliability engineering
- Rca — root cause analysis
- Cp — capital project management
- Rcd — reliability centered design

ACM — Asset Condition Management
- Aci — asset condition information
- Vib — vibration analysis
- Fa — fluid analysis
- Ut — ultrasound testing
- Ir — infrared thermal imaging
- Mt — motor testing
- Ab — alignment and balancing
- Ndt — non destructive testing
- Lu — machinery lubrication

WEM — Work Execution Management
- Pm — preventive maintenance
- Ps — planning and scheduling
- Odr — operator driven reliability
- Mro — mro-spares management
- De — defect elimination
- Cmms — computerized maintenance management system

Leadership

LER — Leadership for Reliability
- Es — executive sponsorship
- Opx — operational excellence
- Hcm — human capital management
- Cbl — competency based learning
- Int — integrity
- Rj — reliability journey

Business Processes

AM — Asset Management
- Sp — strategy and plans
- Cr — corporate responsibility
- Samp — strategic asset management plan
- Ri — risk management
- Ak — asset knowledge
- Alm — asset lifecycle management
- Dm — decision making
- Pi — performance indicators
- Ci — continuous improvement

Reliabilityweb.com's Asset Management Timeline

Business Needs Analysis → Design → Create/Acquire → Operate / Modify/Upgrade / Maintain → Dispose/Renew → Residual Liabilities

— Asset Lifecycle —

Copyright 2016-2020, Reliabilityweb, Inc. All rights reserved. No part of this graphic may be reproduced or transmitted in any form or by any means without the prior express written consent of Reliabilityweb, Inc. Reliabilityweb.com®, Uptime® and A Reliability Framework and Asset Management System™ are trademarks and registered trademarks of Reliabilityweb, Inc. in the U.S.A. and several other countries.

reliabilityweb.com • maintenance.org • reliabilityleadership.com

Reliabilityweb.com® and Uptime® Magazine Mission: **To make the people we serve safer and more successful.** One way we support this mission is to suggest a reliability system for asset performance management as pictured above. Our use of the Uptime Elements is designed to assist you in categorizing and organizing your own Body of Knowledge (BoK) whether it be through training, articles, books or webinars. Our hope is to make YOU safer and more successful.

ABOUT RELIABILITYWEB.COM

Created in 1999, Reliabilityweb.com provides educational information and peer-to-peer networking opportunities that enable safe and effective reliability and asset management for organizations around the world.

ACTIVITIES INCLUDE:

Reliabilityweb.com® (www.reliabilityweb.com) includes educational articles, tips, video presentations, an industry event calendar and industry news. Updates are available through free email subscriptions and RSS feeds. **Confiabilidad.net** is a mirror site that is available in Spanish at www.confiabilidad.net.

Uptime® Magazine (www.uptimemagazine.com) is a bi-monthly magazine launched in 2005 that is highly prized by the reliability and asset management community. Editions are obtainable in both print and digital.

Reliability Leadership Institute® Conferences and Training Events (www.reliabilityleadership.com) offer events that range from unique, focused-training workshops and seminars to small focused conferences to large industry-wide events, including the International Maintenance Conference (IMC), MaximoWorld and The RELIABILITY Conference™ (TRC).

MRO-Zone Bookstore (www.mro-zone.com) is an online bookstore offering a reliability and asset management focused library of books, DVDs and CDs published by Reliabilityweb.com.

Association of Asset Management Professionals (www.maintenance.org) is a member organization and online community that encourages professional development and certification and supports information exchange and learning with 50,000+ members worldwide.

A Word About Social Good

Reliabilityweb.com is mission-driven to deliver value and social good to the reliability and asset management communities. *Doing good work and making profit is not inconsistent,* and as a result of Reliabilityweb.com's mission-driven focus, financial stability and success has been the outcome. For over a decade, Reliabilityweb.com's positive contributions and commitment to the reliability and asset management communities have been unmatched.

Other Causes

Reliabilityweb.com has financially contributed to include industry associations, such as SMRP, AFE, STLE, ASME and ASTM, and community charities, including the Salvation Army, American Red Cross, Wounded Warrior Project, Paralyzed Veterans of America and the Autism Society of America. In addition, we are proud supporters of our U.S. Troops and first responders who protect our freedoms and way of life. That is only possible by being a for-profit company that pays taxes.

I hope you will get involved with and explore the many resources that are available to you through the Reliabilityweb.com network.

Warmest regards,
Terrence O'Hanlon
CEO, Reliabilityweb.com

Reliabilityweb.com®, Uptime®, The RELIABILITY Conference™, MaximoWorld and Reliability Leadership Institute® are the trademarks or registered trademarks of Reliabilityweb.com and its affiliates in the USA and in several other countries.